Peptide Biosynthesis:
Prohormone Convertases 1/3 and 2

Colloquium Series on Neuropeptides

Editors

Lloyd D. Fricker, Ph.D., Professor, *Department of Molecular Pharmacology, Department of Neuroscience, Albert Einstein College of Medicine, New York*

Lakshmi Devi, Ph.D., Professor, Principal Investigator, Director, Interdisciplinary Training in Drug Abuse, Associate Dean for Academic Enhancement and Mentoring, *Mt. Sinai School of Medicine, New York*

Communication between cells is essential in all multicellular organisms, and even in many unicellular organisms. A variety of molecules are used for cell-cell signaling, including small molecules, proteins, and peptides. The term 'neuropeptide' refers specifically to peptides that function as neurotransmitters, and includes some peptides that also function in the endocrine system as peptide hormones. Neuropeptides represent the largest group of neurotransmitters, with hundreds of biologically active peptides and dozens of neuropeptide receptors known in mammalian systems, and many more peptides and receptors identified in invertebrate systems. In addition, a large number of peptides have been identified but not yet characterized in terms of function. The known functions of neuropeptides include a variety of physiological and behavioral processes such as feeding and body weight regulation, reproduction, anxiety, depression, pain, reward pathways, social behavior, and memory. This series will present the various neuropeptide systems and other aspects of neuropeptides (such as peptide biosynthesis), with individual volumes contributed by experts in the field.

Published titles

(for future titles please see the website, www.morganclaypool.com/page/lifesci)

Peptide Biosynthesis: Prohormone Convertases 1/3 and 2
Akina Hoshino and Iris Lindberg
www.morganclaypool.com

ISBN: 9781615043644 paperback

ISBN: 9781615043651 ebook

DOI: 10.4199/C00050ED1V01Y201112NPE001

A Publication in the

COLLOQUIUM SERIES ON NEUROPEPTIDES

Lecture #1

Series Editors: Lloyd D. Fricker, Albert Einstein College of Medicine, and Lakshmi Devi, Mt. Sinai School of Medicine

Series ISSN Pending

Peptide Biosynthesis:
Prohormone Convertases 1/3 and 2

Akina Hoshino and Iris Lindberg
University of Maryland–Baltimore

COLLOQUIUM SERIES ON NEUROPEPTIDES #1

 MORGAN&CLAYPOOL LIFE SCIENCES

ABSTRACT

The prohormone convertases (PC) 1/3 and 2 are calcium-activated eukaryotic subtilisins with low pH optima which accomplish the limited proteolysis of peptide hormone precursors within neurons and endocrine cells. We review the biochemistry, regulation, and roles of PC1/3 and 2 in disease, with an emphasis on the work published in the last 10 years. In the 20 years since their discovery, a great deal has been learned about their localization and cellular functions. Both PCs share the same four domains: the propeptides perform important roles in controlling activation and targeting; the catalytic domains confer specificity, with PC1/3 possessing a more restricted binding pocket than that of PC2; the P domain is required for expression and contributes to enzymatic properties; and the C-terminal tail assists in proper routing to granules. PC1/3, but not PC2, exists in full-length and C-terminally truncated forms that exhibit different biochemical properties. Both enzymes associate with binding proteins; proSAAS is thought to modulate precursor cleavage by PC1/3, while co-expression of 7B2 is obligatory for the formation of active PC2. New studies have revealed an increasingly important role for PC1/3 polymorphisms and mutations in glucose homeostasis and obesity.

KEYWORDS

prohormone convertase 1, prohormone convertase 1/3, prohormone convertase 2, proSAAS, 7B2, neuropeptides, obesity, prohormone processing, posttranslational processing, secretory pathway

Contents

1. General Introduction to the Prohormone Convertases 1

2. Prohormone Convertase 1/3 3
   2.1 Introduction to Prohormone Convertase 1/3 3
       2.1.1 Unique Features 3
       2.1.2 Evolution 3
       2.1.3 Domain Structure 5
       2.1.4 PC1/3 Binding Protein: proSAAS 6
       2.1.5 Gene Location 8
   2.2 Distribution 8
       2.2.1 Tissue Distribution 8
       2.2.2 Intracellular Distribution 9
       2.2.3 Cell Lines 9
       2.2.4 Development 10
   2.3 Cell Biology and Maturation 14
       2.3.1 PC1/3 Maturation and Posttranslational Modifications 14
       2.3.2 Targeting of PC1/3 16
       2.3.3 Is PC1/3 a Transmembrane Protein? 17
   2.4 Enzymatic Characterization 17
       2.4.1 General Enzymatic Properties 17
       2.4.2 Contribution of Various Domains to Enzymatic Properties 18
       2.4.3 Substrate Specificity 19
   2.5 Regulation of Expression and Activity 24
       2.5.1 Transcriptional and Translational Control 24
       2.5.2 Endogenous Regulators 26

2.5.3   Regulation of PC1/3 Activity by its Chaperone proSAAS.................... 28

2.5.4   Synthetic Inhibitors and Activators.. 28

2.6   Model Systems, Knockouts and Mutants 29

2.6.1   PC1/3 Knockout Mice 29

2.6.2   PC1/3 Asn222Asp Mutant Mouse 29

2.6.3   proSAAS Transgenic Mice.. 30

2.7   PC1/3 as a Therapeutic Target 31

2.7.1   Human Mutations.. 31

2.7.2   Single Nucleotide Polymorphisms (SNP) 33

2.7.3   PC1/3 and Other Diseases 34

2.7.4   proSAAS and Disease Relevance 34

3.   Prohormone Convertase 2 35

3.1   Introduction to Prohormone Convertase 2.. 35

3.1.1   Unique Features... 35

3.1.2   Evolution 35

3.1.3   Domain Structure and Functions 36

3.1.4   The PC2-Binding Protein 7B2 37

3.1.5   Gene Location... 39

3.2   Distribution 39

3.2.1   Tissue Distribution.. 39

3.2.2   Intracellular Distribution 39

3.2.3   Cell Lines 40

3.2.4   Development 40

3.3   Cell Biology and Maturation 40

3.3.1   ProPC2 Maturation... 40

3.3.2   7B2 and proPC2/PC2 Targeting.. 42

3.4   Enzymatic Characterization 43

3.4.1   General Enzymatic Properties... 43

3.4.2   Contribution of Various Domains to Enzymatic Activity 44

3.4.3   Tissue Analysis of PC2 Activity 45

3.4.4   Substrate Specificity 45

3.5   Regulation of Expression and Activity 47

3.5.1   Transcriptional and Translational Regulation.......................... 47

3.5.2   Endogenous Inhibitors 48

       3.5.3 Synthetic Inhibitors and Activators 49

       3.5.4 Regulation of PC2 Activity by its Chaperone 7B2 49

       3.5.5 Regulation of 7B2 Expression 50

  3.6 Model Systems, Knockouts, and Mutants 51

       3.6.1 The PC2 Knockout Mouse 51

       3.6.2 The 7B2 Knockout Mouse 52

       3.6.3 Model Systems 53

  3.7 PC2 as a Therapeutic Target: Disease Relevance 55

       3.7.1 Polymorphisms 55

       3.7.2 Cancer 55

4. **Summary and Future Directions** **57**

References **59**

Author Biography **97**

CHAPTER 1

General Introduction to the Prohormone Convertases

The idea that peptide hormones and proteins are originally synthesized from inactive large precursors was recognized as early as the mid-1960s, but the enzymes which cleave these precursors to generate bioactive species proved elusive for the next two decades. Since then, many laboratories have concentrated their efforts in identifying the enzymes which mediate not only precursor cleavage and basic residue trimming, but also additional posttranslational modifications, such as amidation, sulfation, glycosylation, acetylation, phosphorylation, and octanoylation—modifications that are often required for bioactivity. While there was early recognition that precursor processing begins with cleavage at the pairs of basic residues—usually Lys–Arg and Arg–Arg—that flank the active peptide, it was not until the late 1980s and early 1990s when the first members of the endoproteolytic machinery, the proprotein convertases (PCs) were identified, cloned, and characterized.

The first indications of proteolytic activity able to cleave a peptide hormone precursor were provided by John Hutton and colleagues, who demonstrated proinsulin-cleaving activity by two Ca^{2+}-activated enzymes within insulinoma secretory granule extracts [1, 2]. The insulinoma enzyme requiring mM Ca^{2+} that cleaved proinsulin C-terminally to the $Arg^{31}Arg^{32}$ of the B/C junction was termed the "type-1 proinsulin-converting endopeptidase," whereas the enzyme requiring ten-fold lower Ca^{2+} concentrations that cleaved proinsulin C-terminally to $Lys^{64}Arg^{65}$ at the C/A junction was called the "type-2 proinsulin-converting endopeptidase." Due to abundance issues and technological limitations, these enzymes were not purified from secretory granule extracts at this time, but this group provided the first enzymatic characterization of PCs, demonstrating their calcium dependence and mildly acidic pH optima—conditions that are consistent with the intragranular environment. The first use of fluorogenic substrates for the study of PCs was published using these same insulinoma granule enzyme preparations [3].

In 1990, by exploiting the homology of mammalian enzymes to the yeast enzyme kex2 (also known as kexin and kex2p), Steiner and colleagues provided the cDNA sequence of an enzyme they termed PC2 (for protease candidate 2) from human insulinoma [4, 5]. PC1/3 (also known as PC3, SPC3, and Pcsk1) was cloned soon after in 1991 as the third member of the convertase family by two independent laboratories. This enzyme was termed PC3 by Steiner's group (who obtained the sequence from a mouse pituitary cell line, AtT-20) and PC1 by Seidah's group (who utilized both this cell line as well as mouse insulinoma libraries) [6, 7, 8]. By coincidence, the Hutton numbering system perfectly matches the later nomenclature given to the cloned enzymes since it was later shown that type 1 proinsulin-converting endopeptidase, PC1, and PC3 are all the same protein [9]: the enzyme we now call PC1/3. As for the type-2 endopeptidase, although a certain amount of confusion as to catalytic class is apparent in early work due to the sensitivity of the observed activity to the thiol reagent p-chloromercuribenzoate (pCMB), this enzyme was later shown to correspond to the later-cloned enzyme PC2 [10]. Many laboratories quickly confirmed that transfection of PC1/3 and PC2 cDNAs resulted in the expression of enzymes with the catalytic properties and substrate specificity expected of the Hutton proinsulin processing enzymes (reviewed in [11]).

While this review focuses on the two neuroendocrine prohormone convertases PC1/3 and PC2, it is important to note that other proprotein convertase family members exist. These convertases share homology with the catalytic domain of the bacterial enzyme subtilisin and to the yeast processing enzyme kex2. Seven other members of the PC family have now been identified, including the ubiquitously expressed enzyme furin (also known as SPC1), PACE4 (SPC4), PC7 [12, 13, 14, 15]; PC4 (SPC5), found exclusively in the testes [16, 13, 17]; and two isoforms of PC5/6 (SPC6), soluble PC5A/6A, and membrane-bound (PC5B/6B) expressed in the brain, intestine and adrenal [18, 19]. Additionally, two other members of the family—the ubiquitously expressed SKI-I/S1P [20], and PCSK9 (NARC-1), which is expressed in the liver, kidney, brain, and intestine [21]—do not cleave after basic residues. Collectively, these convertases carry out physiological precursor processing reactions ranging from growth factors to peptide hormones, in both the constitutive and the regulated secretory pathways.

Progress on the two neuroendocrine convertases, PC1/3 and PC2, has been rapid. In the last 20 years, these two enzymes have not only been sequenced via cloning efforts but also characterized both in cell systems and in tissues, overexpressed to milligram amounts, and described enzymatically. Two novel binding proteins, proSAAS for PC1/3 and 7B2 for PC2, have been identified. Much has also been learned as to the many regulatory mechanisms which control enzyme activity, and inhibitors have been identified, both endogenous and synthetic. The number of endogenous substrates cleaved by these two enzymes has continuously expanded, which has permitted better predictions as to the processing of other potential precursors. Lastly, exciting new information has recently emerged describing the relationship of polymorphisms and mutations within the PC1/3 and 2 sequences (and their binding proteins) to disease states. While it is not possible to describe each of the hundreds of papers on these two enzymes in depth, our goal in this review is to enable the reader to locate further references in every area of neuroendocrine-related PC research, from evolution to genetics.

CHAPTER 2

Prohormone Convertase 1/3

2.1 INTRODUCTION TO PROHORMONE CONVERTASE 1/3

2.1.1 Unique Features

PC1/3 is an interesting enzyme in many ways. Its restricted expression positions it as one of the two exclusively neuroendocrine precursor processing enzymes, and its activity in the Golgi apparatus and preference for larger substrates indicate that it is likely to initiate the peptide hormone processing cascade [22, 23]. Unlike other members of the convertase family, PC1/3 undergoes extensive autocatalytic processing at the C-terminus to yield multiple active forms with molecular masses of 87 kDa, 74 kDa, and 66 kDa. Compared to other enzymes in the PC family, PC1/3 exhibits extremely slow kinetics [24], and the most active form of PC1/3—66 kDa—has an additional issue of instability with a short half-life of only half hour to a few hours, depending on experimental conditions [25, 26, 27, 28]. Interestingly, increasing reports indicate that mutations and single nucleotide polymorphisms (SNPs) in the *Pcsk1* gene are associated with obesity, allowing the speculation that PC1/3-directed drugs may one day offer new avenues of therapies for endocrine disorders.

2.1.2 Evolution

PC1/3 belongs to a mammalian family of serine proteases which is structurally related to bacterial subtilisins and the yeast subtilase kex2. The enzymes within the PC family share many similarities in the catalytic domain, suggestive of a common origin. PC1/3 shares around 25% sequence identity to subtilisin in this domain and shares 50% with the yeast subtilase kex2 [6]. Most importantly, the catalytic triad is also conserved across the family and is composed of Asp, His, and Ser, and an Asn in the oxyanion hole (Asp in PC2) [29]. The structure of PC1/3 is shown in Figure 1. In the P domain, all mammalian convertases except PC7, contain a signature ^{518}ArgGlyAsp^{520} sequence, a recognition motif for integrins [30, 31, 32] (discussed further below). However, certain non-mammalian convertases, e.g., *Aplysia* PC1/3 [33], appear to have variations in this sequence.

PC1/3 has now been cloned from multiple species, including humans, rodents, fish, the primitive vertebrate *Amphioxus*, and invertebrates such as *Aplysia* and hydra [29, 33, 34, 35, 36]. These evolutionary studies have shown that PC1/3 is a relatively new member of the convertase

Structure of Human PreProPC1/3

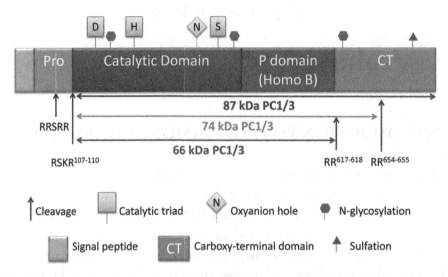

FIGURE 1: Domain structure of preproPC1/3. The designation D, H, S, (purple boxes) represents the catalytic triad Asp, His, and Ser; Asn (N) (orange diamond) is the oxyanion hole residue which stabilizes the transition state. The arrows indicate the cleavage site required for PC1/3 maturation, the pink hexagons represent the three predicted glycosylation sites, and the blue triangle marks the predicted sulfation site. The dashed line in the prodomain indicates the secondary cleavage site, probably cleaved in the Golgi. The P or Homo B domain following the catalytic domain is important for the stabilization of the catalytic domain, as well as determining various enzymatic properties. The C-terminal domain plays a role in efficient routing of PC1/3 to the secretory granules, and contributes to substrate specificity.

family which is most closely related to furin; these two enzymes share 62% of their overall sequence [37] and are 55% identical in their catalytic domains. Due to these similarities, it is possible that few of the PC1/3 sequences found in lower organisms may actually represent furin-like molecules. However, PC1/3s can be distinguished from furins by their lack of the transmembrane segment always found in furin, and their restricted distribution to secretory granules of neurons and endocrine cells. For example, the catalytic domain of hydra PC1/3 is statistically indistinguishable in terms of sequence conservation from that of mouse PC1/3 and human furin, yet this hydra sequence was termed PC1/3 because it lacks a hydrophobic segment following the P domain, making it more PC1/3-like than furin-like [36]. The later evolution of PC1/3 is consistent with its more complex regulation described below, which may relate to the increased complexity of hormonal signaling which evolved in vertebrates.

2.1.3   Domain Structure

PreproPC1/3 is synthesized as a 753-amino acid precursor and consists of the signal peptide and four functional domains. The signal peptide is rapidly removed in the endoplasmic reticulum (ER), resulting in a 94 kDa proPC1/3 zymogen that is comprised of the following four segments: prodomain, catalytic domain, P domain, and C-terminal domain, as shown in Figure 1.

The PC1/3 **prodomain** is 83 amino acids long and is highly conserved (80%) across species. Unlike the catalytic domains, the prodomains of different convertases are not very well conserved (30–40%) across different PC family members, except within a small C-terminal portion [38, 39]. In subtilisin-like enzymes, the prodomain is generally thought to act as an intramolecular chaperone, assisting in the correct folding of the active site [40, 41, 42], though this has not yet been specifically demonstrated for PC1/3. Once folding is established, the prodomain likely remains associated with the enzyme, acting as a potent tight-binding inhibitor with a Ki in the nanomolar range [26, 43, 44, 45].

These two functions of the subtilase prodomain, intramolecular chaperone and inhibitor, have been conserved from bacterial subtilisins [46] to yeast kex2 [47, 48], to modern members of the PC family [41, 42]. PC1/3 forms containing mutations that inhibit prodomain removal result in enzymes that are not properly glycosylated and are retained in the ER [49, 50]. The structure of the mouse PC1/3 prodomain has been resolved by NMR and consists of four anti-parallel β-sheets and two α-helices in a β–α–β–β–α–β arrangement [51].

The **catalytic domain** is highly conserved across the PC family members, with 45% sequence similarity across PC family members and kex2. In PC1/3, this is a 343-residue long domain. Despite the low (26%) homology between the catalytic domains of human PC1/3 and subtilisin, molecular modeling studies have suggested close structural homology, with the exception of a cluster of charged residues within the substrate binding pocket which is most likely responsible for its selectivity for substrates with pairs of basic residues, a feature that is found only in PCs [52, 53]. This domain contains the highly conserved catalytic triad Asp^{167}–His^{208}–Ser^{382}, and an Asn^{309} in the oxyanion hole [38].

The **P domain** (also known as the homo B or middle domain) is the second conserved domain of about 150 residues and is located immediately after the catalytic domain; it is a region unique to the eukaryotic PC family and distinguishes this family from bacterial subtilisins and other enzymes. Molecular modeling studies predicted that the P domain consists of a β-barrel structure with a hydrophobic core; this was later verified when the crystal structure of the related enzyme furin was obtained [54]. The P domain is involved in many aspects of PC1/3 maturation and activity including zymogen activation [31, 55], correct routing of PC1/3 to the secretory granules and overall secretion [31, 56, 57]. The P domain regulates enzymatic activity by assisting in the folding and stabilization of the catalytic domain [57, 58] and determines enzymatic properties, such as calcium binding, low pH dependence, and substrate specificity [58].

The P domain contains several key residues that are well conserved between species. Firstly, residues Gly^{593} and Thr^{594} represent the functional boundary of this domain and play an important role in overall enzyme stability [59]. In particular, Thr^{594}, which cannot be replaced by Ser, is required for proper catalytic domain folding [57]. Secondly, the sequence ^{517}ArgArgGlyAsp^{520} (RRGD; numbering using the mPC1/3 sequence) is found in six of the nine PCs and is conserved in most species that possess PCs [31]. The RRGD motif is critical for proPC1/3 processing and correct routing of 87 kDa PC1/3 to the secretory granules [31, 56]. In *in vitro* studies, PC1/3-derived peptides containing the RRGD sequence bind to the integrin α5β1; however, *in vivo*, interaction of PC1/3 with integrin is only observed in the ER and does not require the RGD sequence. Therefore, the biological relevance of the integrin binding motif is unclear as yet.

The **C-terminal domain** is the least conserved domain between different members of the convertase family, varying both in length and in sequence. PC1/3 has an exceptionally long C-terminal domain of ~21 kDa. Several studies have shown that the C-terminal tail is involved in the proper sorting and routing of PC1/3 to the secretory granules [31, 50, 56, 60, 61, 62, 63, 64, 65]; it is removed once PC1/3 enters the secretory granule [27, 66, 67, 68, 69, 70] (discussed further below).

2.1.4 PC1/3 Binding Protein: proSAAS

proSAAS was the second neuroendocrine protein to be characterized as a convertase binding partner (following 7B2, the binding protein for PC2; see following chapter). ProSAAS was discovered during the mass spectroscopic analysis of brain peptides that were incompletely processed in Cpe^{fat}/Cpe^{fat} mice, which lack active carboxypeptidase E (CPE). Five peptides were found with basic residue extensions at the C-terminus that were all derived from one precursor protein, which was named proSAAS [71, 72]. Full-length 26 kDa proSAAS consists of a proline-rich N-terminal domain and a 41-residue C-terminal tail, which inhibits PC1/3, separated from the N-terminal domain by the conserved furin cleavage site Arg–Arg–Leu–Arg–Arg, as depicted in Figure 2. This domain architecture is quite similar to that of 7B2. ProSAAS is a pan-neuronal protein which exhibits unique properties such as calcium binding and acidic pI (reviewed in [73]). Mammalian proSAAS sequences share over 85% conservation [74] and despite quite limited overall sequence homology (30%) in lower vertebrates such as frog and zebrafish, certain potential alpha helices, basic cleavage sites, and the C-terminal PC1/3 inhibitor region have all been conserved in these lower vertebrates, implying functional attributes [74]. ProSAAS has not yet been identified in any invertebrate, though PC1/3 is clearly present in invertebrate species; lack of identification may again be due to insufficient sequence homology.

The carboxy-terminal 41-residue domain of proSAAS (also known as Big PEN-LEN [75, 76, 77]) is a slow, tight-binding competitive inhibitor of PC1/3 and inhibits this enzyme in the

Structure of Mammalian ProSAAS

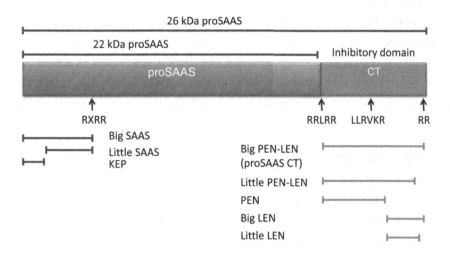

FIGURE 2: Domain structure of proSAAS. The RRLRR furin cleavage site, and other processing sites are indicated by an arrow. Peptide products derived from the N-terminus of proSAAS are shown in blue, and peptide products derived from the C-terminus are in green. The inhibitory hexapeptide, LLRVKR, is shown. 26 kDa and 22 kDa proSAAS have also been referred to as 27 kDa and 21 kDa proSAAS, respectively, by other laboratories.

nanomolar range with limited specificity [76, 71, 75]. The kinetics of inhibition are similar to those of the 7B2 carboxy-terminal peptide for its cognate enzyme, PC2. The inhibitory region can be narrowed to a hexapeptide—Leu–Leu–Arg–Val–Lys–Arg—which is found within the PEN product [75, 78]. Binding assays indicate that proSAAS binds to 71 kDa PC1/3 and to a lesser degree to 74 kDa PC1/3, but does not bind to 85 kDa PC1/3 (variations in the reported molecular weights of PC1/3 are due to different enzyme sources) [75].

Although this review is focused mainly on proSAAS as it relates to PC1/3 activity, it is important to mention that additional functions are now emerging for the various proSAAS-derived peptides. ProSAAS exhibits a much broader distribution pattern than that of PC1/3 and has recently been shown to participate in a number of physiologically important systems including circadian rhythm [79], energy balance [80], food intake [81], and fetal neuropeptide processing [80]. Finally, proSAAS has been found in tau inclusions [82, 83], is implicated as a potential biomarker for neurodegenerative diseases [84, 85, 86], and serves as a potent anti-aggregation chaperone for $A\beta_{1-42}$ and α-synuclein (A. Hoshino and M. Helwig *et al.*, in preparation).

2.1.5 Gene Location

The *Pcsk1* gene which encodes PC1/3 is located on chromosome 5q15–21 in humans and consists of 14 exons, covering about 43.8 kb [87]. Northern blot analysis revealed a dominant 6.2 kb mRNA transcript in humans [8]. In mice, the *Pcsk1* gene is found on chromosome 13c and consists of 15 exons and 14 introns that span at least 42 kb. In tissues, two mRNA transcripts of 2.8 and 4.4 kb are found which result from the use of two different polyadenylation sites located in exons 14 and 15 [7, 88]. *Pcsk1* has also been localized to rat chromosome 2q1 [89, 90]. The gene *Pcsk1n* encoding proSAAS is located on mouse chromosome XA1.1, and on the human chromosome Xq11.23, and contains 3 exons.

2.2   DISTRIBUTION

2.2.1   Tissue Distribution

In situ hybridization and immunohistochemical studies have demonstrated that PC1/3 expression is highest in neuroendocrine and endocrine tissues [91, 92, 93, 94, 95, 96]. In the brain, PC1/3 expression is richest in the hypothalamus, specifically in the magnocellular neurons within the paraventricular and supraoptic nuclei; this is consistent with the ample presence of neuropeptide precursors in these regions [97]. Outside the hypothalamus, PC1/3 is found in layers IV and V of the cerebral cortex; in the pyramidal cell layer and stratum oriens within the CA1–CA3 regions in the hippocampus; in thalamus; in larger neurons such as the interneurons in the striatum; in Purkinje and granular cells of the cerebellum; along the border of corpus callosum; and in the dorsal horn of the spinal cord [7, 97, 98, 92, 99]. Expression extends to retinal neuronal cells, where PC1/3 mRNA is expressed more strongly in the retina than in the optic nerve head; the PC1/3 protein is specifically found in the ganglion cell layer and in the nerve fiber layer of the inner retina [100].

In peripheral tissues, PC1/3 is present in all three lobes of the pituitary [7, 93, 99, 101], the thyroid, the adrenal medulla, in intestinal L cells, and in differentiating epithelial cells in the colorectal zone and gut [94, 99, 102, 103, 104, 105, 106]. In the pancreas, the highest levels of PC1/3 are found in β-cells, although PC1/3 is expressed in adult α-cells at low levels [103, 107, 108, 109]. Additionally, PC1/3 has been found in non-neuronal tissue, possibly as part of local opioid neuropeptide networks recently identified in the gut, spleen, skin, and lung [110, 111, 112, 113]. PC1/3 is also expressed in lung tumors, within alveolar macrophages, submucosal glands, cancerous cells infiltrating bronchial lesions, and neuroendocrine cells and nerves of the bronchial epithelium, but no PC1/3 has been detected in healthy lung tissue [113, 114, 115, 116]. PC1/3 is additionally found in spleen mononuclear cells, splenic macrophages, human monocyte-derived macrophages, and human monocytic leukemia cells [116, 117]. PC1/3 is not detectable in thyroid follicular cells, parathyroid, adrenal cortex, testis, anal canal, undifferentiated monocytes, lymphocytes, or lymph follicles [92, 94, 117, 118].

As discussed above, proSAAS and proSAAS-derived peptides exhibit a broader distribution pattern than PC1/3, but are still largely restricted to neural and endocrine tissues [99]. The highest expression of proSAAS and proSAAS processing products occurs in the hypothalamus and includes the arcuate nucleus, supraoptic and superchiasmatic nuclei [81, 92, 119, 120, 121, 122]. ProSAAS is also found in the cortex, dentate gyrus and CA1-3 of the hippocampus, piriform cortex, Purkinje cells in the cerebellum, and the pontine nucleus.

In the periphery, proSAAS is found in all lobes of the pituitary; co-localization with PC1/3 was demonstrated in melanotrophs [99, 119, 120]. In the adrenals, proSAAS co-localizes with PC1/3 in the adrenal medulla, but proSAAS immunoreactivity can also be found in the outer edges of the cortex where aldosterone biosynthesis takes place [99]. Interestingly, in the pancreas, proSAAS-derived peptides are richest in alpha and delta cells rather than in β-cells, the major pancreatic source of PC1/3 [123, 124]. In the intestine, proSAAS is found predominantly in the external layers and in the outer edges of the small intestine (as opposed to the internal intestinal epithelium layer) [92]. Finally, while PC1/3 is not detectable in the testis, proSAAS mRNA is found in the Leydig cells of the testis [92].

In summary, the distribution patterns of PC1/3 and proSAAS, while highly correlated, do not totally overlap, since proSAAS expression is also observed in non-PC1/3 immunoreactive cells. The differences in proSAAS and PC1/3 expression support the hypothesis that proSAAS plays multiple roles *in vivo* which include regulation of PC1/3 activity, but also include functions unrelated to PC1/3 and neuropeptide production.

2.2.2   Intracellular Distribution

In AtT-20 cells, PC1/3 is found in the trans-Golgi network (TGN) as well as the cell tips, which contain the dense core secretory granules [102, 125, 126, 127]. 87 kDa PC1/3 is the predominant species in the TGN, where it is thought to participate in limited proteolysis of precursors before being sorting to granules and converted to 66 kDa PC1/3 [68, 125, 127, 128]. How PC1/3 is targeted to secretory granules is not fully understood; there is evidence for the involvement of the C-terminal tail domain, discussed further below. It is estimated that PC1/3 and proopiomelanocortin (POMC) are found in a 1:5 ratio in AtT-20 cell secretory granules [67].

2.2.3   Cell Lines

Tumor cell lines derived from endocrine tissues often express PC1/3 [92, 94, 101, 118]. The most commonly used cell lines which express PC1/3 include the mouse pituitary line AtT-20 and the pancreatic beta cell line β-TC6 [129]. NIT2 and NIT3 cells are two β-cell lines derived from CPE$^{fat/fat}$ (mutant CPE) mouse pancreas [130]. Both cell lines lack CPE activity; NIT2 cells also lack both

PC1/3 and 2, and NIT1 cells represent control cells that express wild-type CPE and PC1/3 and 2. In contrast, the rat pituitary cell lines GH3 and GH4C1, the mouse pancreatic cell line α-TC6, and the mouse neuroblastoma Neuro2a cells express only PC2. These lines are suitable for studying transfected PC1/3 enzymes because they possess a regulated secretory pathway [66, 91, 131]. GT1-7 (mouse hypothalamic), SK-N-SH, SK-N-MC (human neuroblastoma) cells also express very low or non-detectable levels of PC1/3 [132]. PC1/3 is expressed in rat bone marrow stromal stem cells and THP-1 cells only after differentiation into neurons and macrophages, respectively [117, 133], whereas PC1/3 expression is lost when the NT2 (human neuron-committed teratocarcinoma) cell line is differentiated [134]. CHO cells, mouse L cells, and insect cells have all been used to produce recombinant PC1/3 [24, 25, 26].

ProSAAS is expressed exclusively in endocrine- and neural-derived cell lines, including Neuro2a, the neuroblastoma NG108; the pituitary-derived lines GH4C1, GH3, and AtT-20; pancreatic lines such as Rin5f and β-TC3 and the adrenal line PC12 [92].

2.2.4   Development

The known expression patterns of PC1/3 and proSAAS mRNA and protein during rodent development are summarized in Table 1. PC1/3 is first expressed briefly in both unfertilized eggs and in preimplantation embryos. During this very early stage (E0-E5), PC1/3 is expressed in the cytoplasm, except when it transiently translocates to the pronuclei after fertilization until the post-zygotic stage [135]. Expression of proSAAS begins at E7, and PC1/3 expression returns during early mid-gestation at E9 [136]. In *Xenopus*, expression of proSAAS begins only at the neurula stage; oocytes do not express this protein [74].

As the embryo develops, PC1/3 and proSAAS expression gradually increase in a subset of tissues, which during the final stages of development begin to become restricted to the brain and endocrine tissues, resulting in the distribution observed in adult tissues; this pattern correlates with neuropeptide production. For example, Marcinkiewicz *et al.* found that PC1/3 mRNA and protein can be detected in the anterior lobe from E15 and E17 and that expression continues to increase until maximum level is reached in adulthood [137]. This expression pattern coincides with POMC expression and increasing adrenocorticotropic hormone (ACTH) production, indicating the importance of PC1/3 activity in neuropeptide production beginning very early in life. Differences in expression exist between adult and embryonic tissues: PC1/3 is a neuronal protein in the adult, but during development, it is also found in glial cells in the embryonic brachial plexus [138]. While PC1/3 is expressed throughout the adult pituitary, PC1/3 is rarely detected in the intermediate lobe during prenatal and early postnatal stages [138]. Finally, proSAAS also plays an important role in neuropeptide production. At E15.5, as processing of proSAAS initiates and the inhibitory proSAAS peptide PEN-LEN

TABLE 1: Timeline of PC1/3 and proSAAS development in mice and rats.

AGE (MOUSE DEVELOPMENT)	EXPRESSION	TISSUE	REFERENCE
Unfertilized eggs	PC1/3 mRNA and protein	Cytoplasm	[135]
Fertilized (E1)	PC1/3	Pronuclei	[135]
Post-zygotic (≈E4.5)	PC1/3	Cytoplasm, junction between blastomeres	[135]
E7	ProSAAS	Diffuse distribution	[136]
E9	PC1/3 and proSAAS	Widely expressed especially in neural tube, hindbrain, telencephalon	[136]
E11	proSAAS	Mid-brain	[136]
	proSAAS, weak PC1/3-ir	First branchial arch mesoderm, liver	[136]
E13–17	PC1/3 and proSAAS	Brain	[136]
	PC1/3 and proSAAS begin to decrease	Non-endocrine tissue	[136]
E15	PC1/3 mRNA	Anterior pituitary	[137]
	proSAAS	Dorsal root ganglion	[136]
E15.5	proSAAS (SAAS, PEN-LEN isoforms)	Brain	[77]

TABLE 1: (*continued*)

AGE (MOUSE DEVELOPMENT)	EXPRESSION	TISSUE	REFERENCE
E15–E17	P1/3 and proSAAS	Proliferating cells in the ventricular zone, cortical plate, trigeminal ganglion	[136]
E16	PC1/3 and ACTH co-localization	Anterior pituitary (Corticotrophs)	[137]
	PC1/3	Anterior lobe (Gonadotrophs)	
	PC1/3 and proSAAS	Pancreas, dorsal root ganglion	[136]
E18	PC1/3 mRNA begins to increase until adulthood	Anterior pituitary	[137]
	PC1/3 mRNA + protein	Intermediate pituitary	[137]
P1	PC1/3 mRNA (maximum at P42)	Neurointermediate pituitary	[137]
P3–adult	PC1/3 protein	Nerve terminals of pituitary posterior lobe	[137]

AGE (RAT DEVELOPMENT)	TISSUE	EXPRESSION	REFERENCE
E12	Pancreas, intestine, brain, lung	PC1/3 mRNA	[422]
	pancreas	PC1/3 and glucagon	
E13	Developing nervous system; anterior pituitary	PC1/3 mRNA	[138]
E14.5	Postoptic area of the hypothalamus, ventral region of subthalamus, subregion of hindbrain, cranial ganglia, dorsal root ganglia, sympathetic ganglia	PC1/3 mRNA	[138]
E16	Cranial and spinal ganglia, branchial plexus	PC1/3 mRNA	[138]
E18	Pancreas	PC1/3 mRNA	[138]
P1–P15	Anterior pituitary but not in intermediate pituitary	PC1/3 mRNA and POMC	[138]

mRNA data were obtained from RT-PCR or *in situ* hybridization experiments ("mRNA"). The days indicate the first day on which the protein was first detected in the tissue.

begins to accumulate, no prodynorphin processing is observed. However, prodynorphin processing is initiated once PEN-LEN levels become undetectable in the adult brain extracts [77]. Overall, the C-terminus of proSAAS shares a similar developmental distribution with PC1/3, in most regions preceding expression by about 2 days. These studies point to an important role for proSAAS in regulating neuropeptide processing during development.

2.3   CELL BIOLOGY AND MATURATION
2.3.1   PC1/3 Maturation and Posttranslational Modifications

PC1/3 undergoes extensive posttranslational modifications which begin in the ER and continue as PC1/3 transits through the regulated secretory pathway (Figure 3). In the ER, PC1/3 is ini-

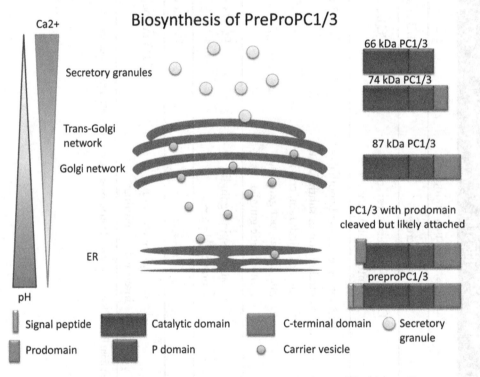

FIGURE 3: PC1/3 maturation during the regulated secretory pathway. This figure illustrates the process of PC1/3 maturation as it moves through the regulated secretory pathway. In the ER, PC1/3 is synthesized as preproPC1/3 and the signal peptide and the prodomain are both rapidly cleaved. The cleaved prodomain most likely remains associated until 87 kDa PC1/3 reaches the Golgi apparatus. In the secretory granules, PC1/3 is C-terminally truncated, yielding the 74 kDa and 66 kDa forms of PC1/3. The color scheme is the same as in Figure 1.

tially synthesized as a 97 kDa proPC1/3 species, but within minutes of zymogen synthesis—$t_{1/2}$ of 2 minutes—the prodomain is rapidly cleaved [69, 139]. There are three potential cleavage sites within the prodomain: ^{50}RRSRR^{54}, ^{61}KR^{62}, and ^{80}RSKR^{83}; the primary cleavage site, which is highly conserved, has been established as ^{80}RSKR^{83}, and this initial cleavage is required for PC1/3 to exit the ER [44, 49, 69]. It is unclear precisely where along the secretory pathway the prodomain dissociates from PC1/3. For furin, the prodomain remains associated until the second cleavage event, and dissociation requires a decrease in pH in the TGN, which triggers His protonation and a conformational switch [140]. This could also be the case for PC1/3, since PC1/3 activity is not detected until this protein reaches the TGN [128]. If PC1/3 activation resembles that of furin, this propeptide histidine switch would add another layer of regulation to block premature activation of PC1/3.

Amazingly, the prodomains of PC1/3 and furin share only about 30% homology and yet can be swapped without affecting the rate of PC1/3 secretion [50]. PACE4/PC1/3 prodomain chimeras undergo efficient prodomain removal, while switching prodomains between PC1/3 and PC2 results in lack of prodomain cleavage and ER retention [50]. A PC1/3 construct lacking the P domain also cannot remove its prodomain [57, 141]; this is not specific to PC1/3, as deletion of the furin P domain also leads to lack of furin activity [142]. Mutation of the catalytic domain supports the idea that the prodomain removal is an autoproteolytic and intramolecular process; as mentioned above, proteins containing mutations in the catalytic triad that block proPC1/3 conversion to mature PC1/3 are retained in the ER, and addition of active enzyme *in trans* cannot rescue prodomain cleavage [49, 50]. However, mutation of the oxyanion hole to Asp—as found in PC2—does not result in a dramatic decrease in PC1/3 prodomain removal [143]. Finally, under pulse/chase conditions, expression of the propeptide in AtT-20 cells inhibits 66 kDa PC1/3 formation, and consequently reduces POMC processing [144]. In HEK cells, *in trans* expression of the prodomain significantly decreases processing of the zymogen [144].

PC1/3 undergoes several additional posttranslational processing events during biosynthesis, for example, N-glycosylation and sulfation [26, 139, 145]. There are three potential glycosylation sites, of which at least two are used, and glycosylation of Asn^{146} is especially critical for proper folding and prosegment cleavage [7, 129, 145, 146, 147]. In the Golgi, terminal glycosylation takes place, as indicated by the acquisition of resistance to endoglycosidase H [139, 145]. Zandberg *et al.* found that inhibiting PC1/3 glycosylation altered POMC processing, suggesting that this post-translational modification might play a role in regulating PC1/3 activity [147]. Sulfation is thought to occur on a specific Tyr in the C-terminal domain and not on the glycosylation sites [26, 145]. Although this has not yet been experimentally demonstrated, by analogy with furin, a second internal cleavage within the prodomain may also occur in the Golgi, resulting in dissociation of the proregion and generating active 87 kDa PC1/3. PC1/3-mediated POMC processing has been observed in the TGN [67, 69, 148, 149], suggesting that precursor processing is initiated in this subcellular compartment.

The final processing of the PC1/3 molecule occurs in the immature secretory granules, which produces an intermediate 74 kDa form and the most active 66 kDa PC1/3 species [27, 66, 67, 68, 69, 70]. Although many dibasic cleavage sites are present in the C-terminal tail, Arg–Arg[617-618] represents the actual *in vivo* cleavage site [27, 62, 150]. The resulting C-terminally truncated 66 kDa PC1/3 is much more active than 87 kDa PC1/3, is highly unstable, and has a lower and narrower pH optimum [26, 27, 151]; this pH is appropriate to the secretory granule environment.

2.3.2   Targeting of PC1/3

Many peptide hormone precursors are processed to bioactive peptides only after arrival at the dense core granules; therefore, it is critical for both prohormones and processing enzymes to be correctly and efficiently routed. Selective calcium-dependent aggregation into granules, as well as interaction with sorting receptors, have both been proposed as possible mechanisms to direct proteins into granules (reviewed in [152]). However, the process by which secretory proteins is sorted into the regulated pathway (in which secretory proteins are stored in granules until a stimulus arrives) versus the default constitutive pathway (in which secretory proteins are continuously trafficked to the exterior) is not yet fully understood.

The C-terminal domain of PC1/3 clearly plays a role in routing PC1/3 to secretory granules in AtT-20, PC12, and GHC4 cells [31, 56, 50, 60, 61, 62, 63, 64, 65]. This domain is highly acidic and contains two α-helical structures which when bound to Ca^{2+} act to sort PC1/3 into granules [61, 63, 153]. Residues 617–753 target PC1/3 to the secretory granules and the following segments, residues 617–625, 665–682, and 711–753, assist with sorting and association with membranes [61, 65]. However, since membrane association of both forms—66 kDa and 87 kDa PC1/3—has been reported [61, 154, 155, 156, 65, 70], the C-terminal tail cannot be solely responsible for membrane association of this enzyme.

Fusion of the PC1/3 C-terminal tail to reporter proteins can re-route an otherwise constitutively secreted protein to secretory granules [153, 157]. The addition of an α-helical segment from the PC1/3 C-terminus together with other secretory granule sorting signals enhances the routing of these proteins to secretory granules, suggesting that not only is the tail important for the sorting of PC1/3 but that it can also modulate the sorting efficiency of prohormones [153, 157]. Conversely, the majority of 66 kDa PC1/3 artificially expressed without a C-terminal tail does not enter granules [50] (but see also [150] where no effect of the C-terminal tail was observed), and disruption of the tail α-helix results in a PC1/3 molecule that is no longer secreted via the regulated pathway [63]. These results highlight the importance of the C-terminal domain and the tail α-helices in targeting

PC1/3 to the secretory granule. Interestingly, chimeras of the PC1/3 prodomain and C-terminal domain are not directed to secretory granules, most likely because they are processed in the ER by furin or PC7 [61, 158].

2.3.3   Is PC1/3 a Transmembrane Protein?

PC1/3 has been reported to be both a Type I transmembrane protein and a peripheral membrane-associated protein. Several groups have found that residues 619–638 associate with lipid rafts in the membrane [65, 155, 156]; the Loh group has found that residues 727-751, encoding a potential α-helix, interact with the membrane on the cytosolic side [156]. However, various biochemical assays, such as phosphorylation, glycosylation, and protease protection assays, clearly demonstrate that PC1/3 is neither a transmembrane protein nor initially synthesized as a transmembrane protein in the ER [159]. These authors also point out that the proposed transmembrane sequence is unusual and would be energetically highly unfavorable for membrane integration. In addition, many other laboratories have found 87 kDa PC1/3 in secretion media; this would be unlikely to occur if this form of PC1/3 were a true transmembrane protein [160, 24, 26, 25, 161]. We conclude that while PC1/3 clearly associates strongly with membranes, it is a peripheral rather than an integral membrane protein.

2.4   ENZYMATIC CHARACTERIZATION
2.4.1   General Enzymatic Properties

Several laboratories have successfully purified recombinant PC1/3, but recombinant enzymes produced from all systems exhibit very low specific activity against both fluorogenic and natural substrates [24, 25, 26, 161, 162]. Compared to furin, which processes the fluorogenic substrate pERTKR-aminomethylcoumarin (amc) at rates of 11–30 μmol amc/mg/h [163, 164, 165], and PC2, with rates of 17–29 μmol amc/mg/h [166, 167], PC1/3 is extremely slow, exhibiting a rate of only 134–480 nmol amc/mg/h [24, 25]. The C-terminally truncated 66 kDa form of PC1/3, either generated as a truncated recombinant protein, or obtained via cleavage of 87 kDa PC1/3, exhibits a 4- to 40-fold increase in specific activity both *in vitro* and *in vivo*, but is highly unstable with a half-life of only 30 minutes to a few hours [26, 27, 151, 25, 28].

87 kDa PC1/3 has a broad pH optimum of 5–6.5, consistent with the environment it experiences in the Golgi and in secretory granules [24, 126, 158]. The generation of 66 kDa PC1/3 occurs spontaneously in an intermolecular fashion, is PC1/3 concentration-dependent, and requires a high Ca^{2+} and acidic environment [26, 27]. While both PC1/3 forms require Ca^{2+} for activity, C-terminally truncated PC1/3 exhibits a 15-fold higher Ca^{2+} requirement and a lower, narrower pH optimum of 5-5.5 for maximal activity [27]. The removal of the C-terminal tail also alters substrate specificity. For example, 87 kDa PC1/3 cleaves provasopressin only at the neurophysin/

glycopeptide junction, whereas 66 kDa PC1/3 will cleave at the neurophysin/glycopeptide and at the vasopressin/neurophysin junction, producing three products. Another example is prooxytoxin, which is cleaved only by 74/66 kDa PC1/3 to release oxytocin and neurophysin [168]. Recently, we have proposed that a PC1/3 dimer represents a novel active form of PC1/3, but further work is required to characterize the enzymatic properties of this form [28].

PC1/3 can be assayed *in vitro* using pERTKR-amc at 200 μM final concentration, in sodium acetate or MES buffer (0.1 M, pH 5.5–6) containing 2–5 mM calcium and detergent, either 0.1% Brij35 or 0.4% octylglucoside. The presence of BSA at 0.1 mg/ml can enhance activity by blocking enzyme adsorption to tube walls. The K_m of PC1/3 for this fluorogenic substrate is 11 μM [169]. An assay for this enzyme in tissue extracts has recently been described [170] which depends on the use of nonspecific protease inhibitors in the initial homogenate, then measuring enzyme activity in the presence and absence of micromolar concentrations of Leu–Leu–Arg–Val–Lys–Arg–NH$_2$ to yield the specific contribution of PC1/3 to total hydrolysis.

2.4.2 Contribution of Various Domains to Enzymatic Properties

Short peptides within the prodomain of PC1/3 (residues 83–93 surrounding the cleavage site between the prodomain-catalytic domains) have been used as an inhibitor platform. Modifications with single amino acid substitution, ketomethylene pseudopeptide bond or replacement of the P1' position with unnatural amino acids such as γ-aminobutyric acid (GABA), beta-Cha, beta-Ala, L-Tic, yield more specific, micromolar inhibitors of PC1/3 [44, 171, 172, 173, 174]. However, these peptide inhibitors are still susceptible to inactivation by proteolytic cleavage and display relatively poor membrane penetrance. Basak *et al.* have also experimented with natural compounds purified from a popular Asian medicinal plant, *Andrographis paniculata*. Modifications to the major constituent of the plant, succinylated andrographolide-based labdane diterpines, display micromolar Kis against PC1/3, and may represent a new source of PC1/3 inhibitors [175]. Unfortunately, these inhibitors are also excellent inhibitors of furin, and thus lack specificity.

Domain swapping and mutagenesis experiments permit us to gain insight into how each domain contributes to enzymatic properties. To address the role of PC1/3-specific sequences within the catalytic domain to enzymatic properties, Ozawa *et al.* switched small patches of residues in PC1/3 with those corresponding to PC2 sequences; the expectation was that specific activity could be increased [176]. The activity of 87 kDa PC1/3 was not affected but a shift in pH optimum was observed, indicating that the primary sequence may be less important in determining PC activity than contributions from the different domains. In addition, while the specific activity of 66 kDa PC1/3 benefited from one of these mutations, when the mutant 66 kDa PC1/3 was transfected into PC12 cells, the efficiency of proneurotensin cleavage was unaltered. Clearly, we do not completely understand the regulatory mechanisms of PC1/3 activity *in vivo* [176]. PC1/3 and furin can cleave similar substrates *in vitro* [177] suggesting that PC1/3 and furin may share similar characteristics in the binding pocket, and that differences in substrate specificity may arise at least in part from their differential intracellular localizations.

The P domain is known to contribute to pH optimum and stability [58]; when this domain is swapped with the P domains of either PC2 or furin, the chimeric PC1/3s display 3- to 4-fold higher activity than wild-type PC1/3; oddly, both chimeras exhibit a neutral pH optimum [58], implying that P domain interactions with the active site must play a role in determining enzymatic properties.

The C-terminal tail of PC1/3 clearly plays a role in enzyme stability, as the 66 kDa form of the enzyme lacking this domain is highly unstable. This domain appears to be generally inhibitory, as 66 kDa PC1/3 is by far the most active form of this enzyme, with a 10- to 40-fold increase in activity [26, 27]. In addition, Jutras *et al.* have shown that prorenin processing only occurs in secretory granules after the formation of 66 kDa PC1/3 [60]. *In vitro*, the C-terminal tail exhibits bimodal regulation, acting as an inhibitor at μM concentrations and as an activator at nM concentrations [60, 178]. It is also possible that the tail interacts with the catalytic domain in such a way that it blocks substrate access; in support of this hypothesis, tethering 87 kDa PC1/3 to the membrane via the peptidylglycine α-amidating monooxygenase transmembrane domain increases substrate processing compared to soluble, non-tethered PC1/3, with a broadened pH optimum [179, 180]. All of these studies suggest that interaction between the tail domain and the other domains is critical for determining the enzymatic properties of the 87 kDa form.

2.4.3 Substrate Specificity

PC1/3 most often cleaves neuropeptide precursors and prohormones C-terminally to a pair of basic amino acids. Lys–Arg is the most preferred (50%) dibasic site, although cleavage following Arg–Arg, Arg–Lys, Lys–Lys or after single Arg is also frequently observed [25, 22, 181, 182]. Table 2 depicts a compilation of peptide precursors that are cleaved by PC1/3 and 2; this list expands the one provided in our earlier review [183]. From Table 2, we can deduce that PC1/3 has more restricted substrate specificity than PC2. The readers are referred to a review from Cameron *et al.*, in particular Table V which summarizes various rules governing preferred PC1/3 and 2 cleavage sites as well as inhibitor preferences ([183], but see also later studies [184, 185, 186, 187]). Substrates with Arg at the P4 position are preferred by PC1/3, and substitution at this position with Lys or Orn results in a slight decrease in the k_{cat}/K_m value [171].

For precursors that undergo extensive processing, PC1/3 has a tendency to cleave larger precursors vs. the processing of intermediate-sized neuropeptides mediated by PC2 [183]. Many neuropeptides cleaved by PC1/3 can in fact be cleaved by PC2, but the same cannot be said about PC1/3 processing of PC2 substrates, supporting the idea that the binding cleft is more restricted

TABLE 2: Precursor cleavage by PC1/3 and PC2*.

PROTEIN PRECURSOR	COMMENTS	REFERENCES
Agouti-related protein precursor	Both enzymes, but primarily PC1/3	[423]
Chromogranin A	Both enzymes	[424, 425, 426, 427]**
Chromogranin B	Both enzymes	[428]
Cocaine and Amphetamine-Related Transcript (CART)	Both enzymes	[429]
Histatin-3	PC1/3, PC2 not examined	[430]
Kisspeptin precursor	?	N/a
NESP55	PC1/3, some PC2	[431]
Proadrenomedullin	?	N/a
Proaugurin/egr4	PC2	[314]**
Procalcitonin	PC1/3	[105, 432]
Procholecystokinin	Some sites cleaved by PC1/3; some by both enzymes	[433, 434] [386, 435, 436]**
Procorticotropin-releasing hormone	Conflicting evidence re PC2	[437, 438]
Prodynorphin	Some sites cleaved by both enzymes and some cleaved only by PC2	[22, 343]**

Proenkephalin	Some sites cleaved by both enzymes and some cleaved only by PC2	[439, 440, 438]**
Progalanin	PC2	[438]
Progastrin	PC1/3, Lys–Lys site cleaved by PC2 only	[441, 442, 443, 444]
Proghrelin	PC1/3 only	[445, 177]
Proglucagon	Both enzymes cleave certain sites; others are cleaved by PC1/3 or PC2 only	[382, 446, 447, 448, 449, 450]
Progonadotropin releasing hormone	PC2	**
Proinsulin	B–C junction favored by PC1 while C–A junction favored by PC2	[9, 320, 374, 451, 452]
Proislet amyloid polypeptide	PC2	[401, 453]
Promelanin–concentrating Hormone	Some sites cleaved by both enzymes and some cleaved only by PC2	[454]**
Proneuropeptide W	?	N/a
Proneuropeptide Y	Cleaved more efficiently by PC1/3 than by PC2	[128, 303, 455, 438]**
Proneurotensin/NN	Both enzymes	[456, 457]**
Proopiomelanocortin	Some sites cleaved by both enzymes and some cleaved only by PC2	[67, 148, 341, 458, 459]

TABLE 2: (*continued*)

PROTEIN PRECURSOR	COMMENTS	REFERENCES
Prohypocretin/orexin	?	N/a
Proorphanin/FQ	PC2	[459]
Prooxytocin	PC1/3	[168]
proP518/QRFamide	Both enzymes	(Ozawa and Lindberg, unpublished observations)**
ProPACAP	PC1/3	[460]**
Prorelaxin	PC1/3	[49]
Prorenin	PC1/3	[60, 66]
proSAAS	PC2, PC1/3 slowly	[120]**
Prosomatostatin	Both enzymes	[461, 462, 463, 438]
Prospexin/NPQ	PC2	[464]

Protachykinin A	PC2	**
Protachykinin B	PC2	**
Prothyrotropin-releasing hormone	Some sites cleaved by both enzymes and some cleaved only by PC2	[465, 466, 467, 468]**
Provasoactive intestinal peptide	Not PC2	[438]**
Provasopressin	Conflicting evidence on PC2 enzyme usage at VP-NP junction; PC1/3 can do	[168, 469]
Secretogranin II	Cleaved more efficiently by PC2	[428, 471, 472]**
VGF	PC1/3	[187, 470]**
7B2	PC2, PC1/3 (different sites)	[296, 299]**

*These precursors have been shown to be cleaved by PC1/3 and/or PC2 either in cell culture and/or in vitro.
**indicates data from proteomics studies of PC1/3 and PC2 null mice [184, 185 ,186, 187].
? indicates precursors whose cleavage enzymes are still unknown.

in PC1/3. New studies have allowed the development of improved algorithms to better predict cleavage site usage by PCs (discussed in [188]), but it is clear that we do not yet completely understand the molecular features underlying PC1/3 specificity. Prorenin and proghrelin are examples of unique peptides that are only cleaved by PC1/3 and not by PC2 [60, 66, 177, 445]. Future identification of additional selective cleavage sites will provide information which can hopefully be used to improve our predictions; however, it is likely that tertiary structures will play a part in determining susceptibility to cleavage. Of course, all potential cleavage sites must be experimentally verified in cell lines and/or tissues to verify actual cellular encounter; co-localization of enzymes and substrates must also be shown in tissues.

2.5   REGULATION OF EXPRESSION AND ACTIVITY
2.5.1   Transcriptional and Translational Control

PC1/3 expression is under the control of the following promoter regions: two cAMP-responsive elements (CRE), multiple thyroid response element (TRE)-like sequences, POU sites, and an interferon consensus sequence (ICS). Numerous transcription factors have been implicated in PC1/3 gene regulation, including CREB, ATF-1, Sp1, and c-Jun [88, 189, 190, 191, 192]. Both human and mouse *Pcsk 1* genes contain multiple transcription initiation sites but lack the common TATA or CAAT boxes in promoter regions [88, 190]. The sequence -288 to -1 is responsible for the neuroendocrine tissue-specific expression of PC1/3 [190]. Promoter activity is enhanced by *Pax6* binding in a dose- and Cre-dependent manner, and *Pax6* mutations result in PC1/3 deficiency and abnormal glucose homeostasis [193].

Activation by second messenger pathways is tissue-specific and time-dependent. Long-term treatment of AtT-20 cells with secretagogues, such as 8-bromo-cAMP or phorbol ester, protein kinase A (PKA), and intracellular Ca^{2+} signaling pathways, increase PC1/3 mRNA in AtT-20 cells, gonadotroph adenomas, a human pancreatic carcinoid BON cell line, the neuroepithelioma cell line SKN-MCIXC, and the medullary thyroid carcinoma cell line WE4/2 [194, 195, 196, 196, 197], but not in β-TC3 cells [198]. Activation of protein kinase C stimulates PC1/3 mRNA expression in BON and SKN-MCIXC cells but not in WE4/2 cells [196, 197]. Glucose has been shown to increase the translation of PC1/3 message in pancreatic islets [213]. Taken together, these pathways enable extensive transcriptional and translational control over PC1/3 expression.

Brain PC1/3 gene and protein expression is regulated by many different physiological factors. PC1/3 gene and protein expression is increased in the paraventricular nucleus (PVN) and in the median eminence after leptin administration; in the PVN after induced hypothyroidism by thyroidectomy or 6-n-propyl-2-thiouracil (PTU) treatment; and in the supraoptic nucleus after chronic salt loading [199, 200, 201]. Increased co-localization of PC1/3 and pro-TRH is observed in the PVN after PTU treatment [201]. In contrast, 9-cis-retinoic acid, induced hyperthyroidism, and starvation inhibit PC1/3 promoter activity in the PVN [200]. Morphine also has a time-dependent effect on PC1/3 expression; short-term exposure to morphine down regulates, while long-term exposure to morphine upregulates PC1/3 gene and protein expression in the hypothalamus, in a CRE-dependent manner [202]. Gonadotropin-releasing hormone increases PC1/3 mRNA expression in gonadotroph adenomas [195]. Studies of nerve injury models using kainic acid, pilocarpine, and transected sciatic nerve report upregulation of PC1/3 mRNA and protein in the hippocampus and in Schwann cells [203, 204]. PC1/3 mRNA is not normally found in Schwann cells; surprisingly, it has a similar induction timeline as BDNF after nerve injury [205]. Extracellular matrix proteins including laminin and fibronectin can also enhance PC1/3 levels in neuronal cultures [206].

Interestingly, *ob/ob* mice, which have a mutation in the leptin gene, exhibit differential PC1/3 mRNA expression in the hypothalamus, while PC2 and PACE4 expression patterns are not affected [207]. Compared to the wild-type control, PC1/3 mRNA expression is reduced in the lateral hypothalamus (LH) and increased in the ventromedial nucleus (VMN) [207]. Food deprivation causes mRNA level to decrease in the VMN and to increase in the LH, while leptin administration causes PC1/3 mRNA expression to decrease in the arcuate nucleus and increase in the LH. However, Nilaweera *et al.* conclude that not all of these differences are due solely to leptin [207]; additional work investigating other downstream signaling factors, PC1/3 protein levels, and PC1/3 activity levels will be helpful in understanding the functional effects of the different expression patterns.

In the pituitary, dopamine, thyroid hormones, and corticosteroids play a regulatory role in PC1/3 expression. Haloperidol (a dopamine antagonist) increases and bromocriptine (a dopamine agonist) decreases PC1/3 and POMC mRNA expression in the neurointermediate lobe [129, 208, 93]. In the anterior pituitary, hypothyroidism stimulates PC1/3 expression while hyperthyroidism decreases PC1/3 expression, likely through interaction with the promoter in the TRE-like sequences [93, 189]. In AtT-20 cells, dexamethasone, corticotropic releasing hormone, and glucocorticoid treatment alter PC1/3 expression [93, 129], while lipopolysaccharide (LPS) administration and cytokine treatment (IL-6 and LIF) promote POMC and PC1/3 expression and POMC processing [209]. LPS also increases PC1/3 and proenkephalin levels [116]. In HP75 (human pituitary adenoma) cells, transforming growth factor β1 stimulates PC1/3 mRNA expression [195, 210]. Additional studies will be needed to elucidate the molecular mechanisms for changes in PC1/3 expression levels, but it is clear that PC1/3 plays a pivotal role in responding to environmental and hormonal changes.

Several studies demonstrate PC1/3 gene regulation in the pancreas, especially in adult α-cells which predominantly express PC2 and very low levels of PC1/3 [107, 211, 108]. Exposure to high glucose (25 mM), or β-cell loss as a result of mutation or streptozotocin treatment, can switch expression patterns and upregulate PC1/3 mRNA, altering the peptide products secreted by α-cells [107, 108, 212, 213, 214, 215, 216, 217]. A small molecule agonist for the GPCR TGR5, implicated in GLP-1 release in intestinal L-cells, also increases PC1/3 promoter activity and GLP-1 secretion from α-cells [108]. Finally, during development, pregnancy, α-cell regeneration, and in mouse models of β-cell proliferation, PC1/3 expression is similarly elevated in α-cells [217].

In the intestine, PC1/3 mRNA and protein are up regulated together with proglucagon gene expression in response to GLP-1 secretion, in a CRE- and PKA-dependent manner [218]. Berberine, an anti-diabetic drug, also mildly increases PC1/3 mRNA expression in the enteroendocrine cell line NCI-H716 [219].

There are far fewer studies on the regulation of proSAAS than on PC1/3, but a small number of reports have shown that proSAAS expression responds to environmental stressors such as dehydration [220], hypoxia [221], cold (increased in female hypothalamus) [222], and high-fat diet (increased PEN peptide levels in the female hippocampus) [222]. Mzhavia and colleagues showed that although proSAAS is an endogenous inhibitor of PC1/3, the two proteins are not always co-regulated, as long-term treatment of AtT-20 cells with secretagogues increases PC1/3 mRNA levels without affecting those of proSAAS [194]. ProSAAS mRNA levels are lower in the mediobasal hypothalamus of Cpe$^{fat/fat}$ mice than in control mice [81] and big LEN is slightly elevated compared to wild-type mouse hypothalamus [223]. In Cpe$^{fat/fat}$ mice, food deprivation increases proSAAS processing intermediates—but not the mature peptide products [81, 224].

2.5.2 Endogenous Regulators

Modulation of PC1/3 activity by its own prodomain and the C-terminal domain has been discussed in the sections above. In addition, PC1/3 activity is regulated by self-association [28]. We have recently shown that secreted and intracellular PC1/3 species exist in multiple-sized forms, including inactive aggregated forms, active oligomers, dimers, and known 87, 74, and 66 kDa monomeric forms [28]. Preparations of recombinant PC1/3 that contain both oligomerized and monomeric PC1/3 exhibit latent activity (Figure 4), and we have suggested that these aggregates can dissociate and activate *in vitro*, increasing the specific activity of the preparation [28]. Monomeric 66 kDa PC1/3 also exhibits a lag phase which is dependent on pH, calcium, and substrate concentration; this was hypothesized to represent a slow response to changes in substrate concentration [225].

Within the secretory granules, other potential endogenous regulators of PC1/3 activity include CPE [226] and catecholamines [170, 227]. CPE$^{fat/fat}$ mice contain a point mutation in the CPE gene that renders the enzyme inactive [228]. Interestingly, PC1/3 expression is decreased in the brains and pituitaries of these mice, and general processing of neuropeptide precursors by prohormone convertases is also impaired [226, 228]. The decrease in PC1/3 activity can be partly attributed to incomplete processing of the 87 kDa form to the more active 66 kDa PC1/3 species [226]. In addition, there may be an accumulation of the C-terminal domains of PC1/3 and of proSAAS that remain inhibitory in the absence of CPE activity [226, 228]. However, in AtT-20 cells, siRNA-mediated CPE reduction had no effect on either PC1/3 expression or ACTH pro-

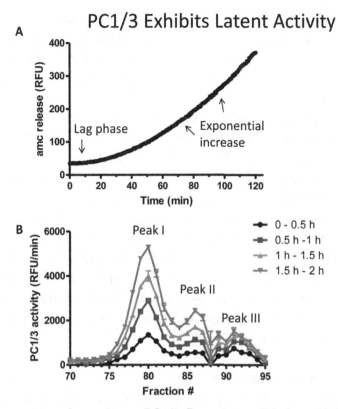

FIGURE 4: Latent activity of recombinant PC1/3. Recombinant PC1/3 purified from CHO/DG44 cells overexpressing mPC1/3 was subjected to ion exchange and gel filtration purification. *Panel A.* PC1/3 exhibits exponentially increasing activity. Fraction 80 from a typical gel filtration chromatography, containing PC1/3 oligomers, was assayed in triplicate against the fluorogenic substrate pERTKR-amc. The activity curve over the course of 2 hours is shown. *Panel B.* Longer assays show increasing PC1/3 activity, especially in larger forms. The activity of fractions 70–95 was assayed. A 2-hour activity assay is here broken into 30-minute intervals to better demonstrate the increasing activity of PC1/3 over time. Peak I contains oligomerized 87 kDa PC1/3, peak II contains 87 kDa PC1/3 dimers and monomers, and peak III contains monomeric 66 kDa PC1/3.

duction in the regulated secretory pathway; it is possible that the siRNA treatment was too short to observe changes in PC1/3 expression and activity [229]. Secretory granules also contain mM concentrations of catecholamines, and treatment of cells with reserpine—a catecholamine uptake inhibitor—results in increased enkephalin levels [230, 231, 232]. We and others have found that this reserpine-induced increase is not due to alterations in enzyme expression levels but to the loss of direct catecholamine inhibition of PC1/3 [170, 227]. Dopamine quinone is the most effective inhibitor among the catecholamines and is irreversible, whereas dopamine, norepinephrine, and epinephrine represent reversible inhibitors of PC1/3 in the mM range [170, 227].

2.5.3   Regulation of PC1/3 Activity by its Chaperone proSAAS

It is now clear that the C-terminus of proSAAS is an endogenous inhibitor of PC1/3 [75, 76]. The inhibition is primarily due to residues 202–207, Leu–Leu–Arg–Val–Lys–Arg, located in the C-terminus of proSAAS (see Figure 2), and mutation and peptide synthesis studies have shown that the Lys–Arg residues are required for inhibition [75, 76, 233, 234]. Interestingly, a proSAAS–GST fusion protein could only be pulled down with 71 kDa PC1/3 and not with the larger 85 kDa PC1/3 protein; association required a mildly acidic pH of 5.5 [75]. *In vitro*, proSAAS is a competitive, slow-binding inhibitor of PC1/3 with a nanomolar Ki ranging from 1.5 to 7.5 nM, depending on the protein/peptide length. ProSAAS fragments can weakly inhibit PC2 as well, but inhibitory potency for PC2 is reduced by 600- to 1500-fold [71, 76]. In AtT-20 cells studied under pulse/chase conditions, overexpression of full-length proSAAS inhibits POMC processing and reduces the rate of conversion of 87 kDa PC1/3 to the 66 kDa form (although under steady-state conditions, POMC processing is not affected) [71, 144, 234]. In HEK cells, co-expression of PC1/3 and the proSAAS C-terminus reduces zymogen processing and the overall secretion of all mature forms [144]. *In vivo*, extensive processing of proSAAS exposes the Lys–Arg of the inhibitory Leu–Leu–Arg–Val–Lys–Arg sequence to CPE, rendering the peptide non-inhibitory; thus, proSAAS appears to only transiently inhibit PC1/3 [119].

2.5.4   Synthetic Inhibitors and Activators

Because of the important role PC1/3 plays in neuropeptide processing, and the wide range of physiological systems this enzyme affects, identifying small molecule modulators of PC1/3 activity would be highly desirable. Several peptide-based inhibitors have been characterized thus far; these have been based either on a Lys–Arg-containing sequence or on the PC1/3 prodomain (reviewed in [235, 236]).

Results from a peptide combinatorial study indicate that peptides containing the sequence Leu–X–Arg–X–Lys–Arg are the most potent inhibitors; indeed the most potent peptide inhibitor in this study, Leu–Leu–Arg–Val–Lys–Arg, has a Ki of 3.2 nM and is the exact inhibitory sequence present within proSAAS [76, 78]. Substitutions at P5 do not affect the Ki except for the deleterious Pro substitution, and Val is favored at the P3 position [78]. Consistent with this finding, decanoyl-Arg–Val–Lys–Arg–chloromethylketone is a widely used irreversible inhibitor of prohormone convertases [237], and peptides containing the Arg–X–Lys–Arg/X–Lys motif are nanomolar inhibitors of PC1/3 [238]. However, these peptides are also prone to cleavage, since they resemble PC substrates. Peptide efficacy can be improved by using other chemical groups to replace the P1 position; a decarboxylated arginine mimetic, phenylacetyl-Arg–Val–Arg-4-amidinobenzylamide, has a Ki of 0.75 μM against furin or PC1/3 [239], and has improved stability. PC1/3 is also inhibited by serpins: wild-type α1-antitrypsin, α1-antitrypsin Pittsburgh, and antithrombin III [25, 161].

We found that short Lys–Arg-containing peptides modulate PC1/3 activity in a time-dependent and enzyme form-dependent manner. At short preincubation periods (min), these peptides inhibit 66 kDa PC1/3, but after longer preincubations (hours), the inhibition is lost and the same peptides instead stabilize PC1/3 activity. In contrast, when 87 kDa PC1/3 dimer-containing preparations are preincubated with these short peptides for 2 hours, enzyme activity is increased almost two-fold (A. Hoshino and I. Lindberg, unpublished observations, [28]). Activation by these Lys–Arg peptides appears to be mediated by both PC1/3 oligomer dissociation and by C-terminal processing (A. Hoshino and I. Lindberg, unpublished observations).

2.6   MODEL SYSTEMS, KNOCKOUTS AND MUTANTS

2.6.1   PC1/3 Knockout Mice

Human and mouse PC1/3s share an overall amino acid homology of 92.6%, with 98% homology in the catalytic domain [8]; therefore, knockout mice represent valuable tools for studying PC1/3 function. Two different PC1/3 knockout mice and one PC1/3 mutant mouse model are now available and exhibit different phenotypes [240, 241, 242]. The first knockout mouse model was generated by targeted deletion of the promoter region and the first exon [240]. We would expect PC1/3 knockout mice to be obese, as humans with PC1/3 mutations exhibit early-onset obesity (see below). Surprisingly, 66% of these homozygous null mutants exhibit neonatal death, and about 33% of the null mice that do survive are runted (Figure 5), due to deficits in PC1/3 processing of pro-growth hormone releasing hormone (proGHRH) and insulin-like growth factor 1 (IGF-1); however, hemizygotes are indeed mildly obese. PC1/3 null mice exhibit deficient processing of other neuropeptide precursors, including POMC, proglucagon, and proinsulin. Not all substrates are affected, implying the presence of compensating redundant processing by other enzymes [184, 187, 243, 244].

The second knockout mouse model has a targeted deletion from exon 2 to exon 10 of the *Pcsk1* gene and displays a more complex phenotype [241]. Homozygous mutants exhibit preimplantation lethality and do not survive past the 8-cell stage, even though 94% of the mutant embryos have the capacity to grow until E6 *in vitro*. Female heterozygote mice that are fed a low-fat diet are somewhat smaller compared to their wild-type littermates, but display age-dependent weight gain when challenged with a high-fat diet. The differences in phenotype between the two null mice are possibly due to genetic background. The first knockout mouse model was generated by injecting 129X1/SvJ-derived embryonic stem cells into C57BL/6J blastocysts, and the resulting mice were crossed to CD1 outbred mice, whereas the second knockout mouse model was generated using C57BL/6J mice and then bred into an FVB/N background.

2.6.2   PC1/3 Asn222Asp Mutant Mouse

Interestingly, mice expressing PC1/3 containing an Asn222Asp mutation mimic the human phenotype and exhibit obesity and hyperphagia which is more pronounced in females [242]. Residue 222

Growth Retardation of PC1/3 Null Mice

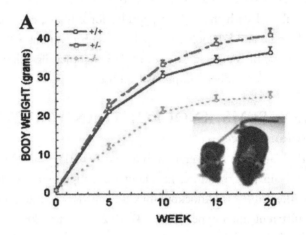

FIGURE 5: PC1/3 KO mice are runted, but not obese. The growth curves of wild-type, heterozygotes and knock-out mice are shown. *Inset:* Instead of being obese, the PC1/3 KO mouse (on the left) is smaller than the WT littermate (on the right) due to impaired processing of growth hormone releasing hormone. *Reproduced with permission from Zhu et al., 2002 PNAS 99.*

in the catalytic domain is highly conserved from bacteria and yeasts to humans, and this mutation results in reduced levels of PC1/3 activity [242, 245]. In addition to obesity, Asn222Asp mice have a defect in insulin production and exhibit β-cell expansion, perhaps to compensate for their reduced insulin; surprisingly, they do not develop diabetes. Heterozygotes display an intermediate weight phenotype unless challenged with a high-fat diet, reminiscent of the second knockout mouse model discussed above.

2.6.3   proSAAS Transgenic and Knockout Mice

With the generation of transgenic mice which overexpress proSAAS and proSAAS knockout mice, we now have multiple novel mouse models to study the function of proSAAS *in vivo* [80, 246]. The first set of mice express a full-length rat proSAAS transgene driven by a beta-actin promoter. These mice exhibit normal pituitary neuropeptide processing and normal body weight until weeks 10–12. At this point, the proSAAS transgenic mice begin to weigh 30–50% more, have twice as much body fat, and exhibit higher elevated blood glucose levels than wild-type mice. However, PC1/3 maturation is not altered in these mice. Since the inhibitory peptide within proSAAS is inactivated by CPE activity, the transgene was also expressed in CPE$^{fat/fat}$ mice in order to test whether proSAAS inhibits PC1/3 *in vivo*. In these mice, the inhibitory effect of proSAAS is more pronounced and processing of pituitary neuropeptides is reduced. Interestingly, these transgenic mice die between weeks 11 and 26 [246].

ProSAAS knockout mice were generated by targeted deletion of transcriptional and translational start sites and exon 1 [80]. These mice are viable but exhibit abnormal neuropeptide processing in the prenatal brain, suggesting that proSAAS inhibits PC1/3 activity *in vivo*. Interestingly, this inhibition is lost in adult animals, and levels of PC1/3 products are restored in adult knockout mice compared to wild-type mice. This result is in agreement with the finding that the inhibitory proSAAS peptide PEN-LEN is not detectable in the adult brain. In addition, proSAAS null mice exhibit decreased locomotor activity, and males weigh 10–15% less than their wild-type controls. The knockout mice are able to maintain normal fasting blood glucose levels and are able to efficiently clear glucose from the blood in response to a glucose challenge.

In summary, these various PC1/3 and proSAAS mutant mouse models highlight the importance of PC1/3 activity in neuropeptide precursor processing and support the hypothesis that at least during development, proSAAS acts an endogenous inhibitor of PC1/3. The first PC1/3 knockout mouse and PC1/3 Asn222Asp mutant mice are available from the Jackson Laboratories. The variations in the phenotypes observed in the different mouse models and in humans (which will be further described in the next section) suggest that genetic background may play an important role in the observed phenotypes.

2.7   PC1/3 AS A THERAPEUTIC TARGET

2.7.1   Human Mutations

Three human cases of compound PC1/3 mutations have been reported so far, and all share the phenotype of early-onset obesity, intestinal dysfunction, and abnormal glucose homeostasis [55, 247, 248]. The three patients have inactivating mutations in both *Pcsk1* alleles and all reported obesity and mild intestinal dysfunction from birth. The first patient, shown in Figure 6, has a Gly593Arg mutation in the P domain which prevents the removal of the prodomain and thus ER exit. This is coupled with a mutation in the second PC1/3 allele encoding a A→C+4 transversion in the donor splice site, resulting in the skipping of exon 5, frameshift and a subsequent premature stop codon in the catalytic domain. Amazingly, this patient was not diagnosed until adulthood, when she was treated for hypogonadotropic hypogonadism. She was able to conceive after treatment with gonadotropin-releasing hormone but developed gestational diabetes mellitus. While impaired plasma POMC and proinsulin processing were observed, processing of other prohormone substrates, such as prorenin and procalcitonin, was not affected. Analysis of plasma from the children of this patient revealed that heterozygosity does not affect PC1/3-mediated insulin biosynthesis [55, 247].

Severe Obesity Resulting from PC1/3 Loss

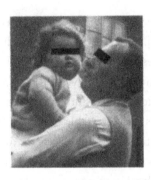

FIGURE 6: Severe obesity resulting from PC1/3 loss. This individual has an inactivating mutation in each of her *Pcsk1* alleles, A-C+4 and a Gly593Arg, resulting in negligible PC1/3 activity. The photo was taken when she was three years old and weighed 80 lbs. *Reproduced with permission from Jackson et al., 1997, Nature Genetics 16: Fig 1.*

The second subject was hypocortisolemic, obese, and exhibited a more severe case of persistent diarrhea caused by malabsorption of monosaccharides and fat, despite a controlled, low-calorie formula diet [247]. She had two mutations in the catalytic domain: the first mutation, Gly250stop, resulted in a truncated PC1/3, and the second, Ala213del resulted in a partially processed enzyme retained in the ER.

The third subject is a 6-year-old boy who is homozygous for Ser307Leu, also in the catalytic domain [248]. Analysis of this mutation in a cell culture model shows that mature PC1/3 is secreted and correctly folded, but does not exhibit any catalytic activity on substrates other than itself. Similar to the previous two cases, this patient exhibited childhood obesity due to hyperphagia and intestinal dysfunction.

The above three cases are examples of patients with two copies of mutant PC1/3 which result in either trace or no PC1/3 activity. Creemers and colleagues recently identified patients heterozygous for 8 different PC1/3 mutations which were either catalytically inactive or underactive [249]. Out of these 8 mutations, one was identified as Asn222Asp, the same mutation that causes obesity in mice [242]. Another mutation was Asn221Asp, a single nucleotide polymorphism (SNP) implicated in human obesity [245, 250, 251]. Interestingly, possessing just one mutant PC1/3 allele increased the risk of obesity by 8.7 times [249].

The mutations discussed above result in a severe loss of PC1/3 enzymatic activity when tested *in vitro*. It is unclear how these mutations produce the wide phenotypic variability observed between patients; the presence of PC1/3 products, despite the lack of PC1/3 activity, indicates the likely contribution of redundant activity by other PCs (such as PC2 and PC5) that compensate for the lack of PC1/3. In addition to metabolism differences, the severe malabsorption observed in these patients suggests that in humans, PC1/3 plays a major role in the normal absorptive function of the small intestine.

2.7.2   Single Nucleotide Polymorphisms (SNPs)

Recent genome-wide association and quantitative trait locus studies have shown that *Pcsk1* polymorphisms are associated with polygenic obesity and an increased risk of Type 2 diabetes [245, 251, 252, 253, 254, 255]. In particular, the nonsynonymous variants rs6232, rs6234, and rs6235 encoding the Asn221Asp, Gln665Glu, Ser690Thr mutations, respectively, have been extensively studied in multiple populations and reproducibly reported to be associated with obesity. Rs6232 (Asn221Asp) is located in exon 6 of the *Pcsk1* gene, has been associated with obesity in multiple European populations, and in one study, shows age-dependency [245, 250, 251]. In HEK293T cells, Asn221Asp PC1/3 exhibits a 10% decrease in activity against the fluorogenic substrate pERTKR-amc [245]. Residue 221 may be important in regulating PC1/3 activity since it is located structurally close to the catalytic residue His^{208}. This residue is also adjacent to Asn^{222}; since Asn222Asp causes obesity in mice [242], a mutation at position 221 may contribute to the obesity phenotype.

Rs6234 (Gln665Glu) and rs6235 (Ser690Thr) are linked nonsynonymous variants located in exon 14, and the resulting mutations are both found within the C-terminal tail of PC1/3. These SNPs are associated with 1.22-fold increase in the risk of obesity (either alone or as a pair) in Danish, Swiss, French, Swedish, and Greek populations [245, 250, 254]. A decrease in postabsorptive, but not postprandial, resting energy expenditure was significant until the data were adjusted for fat mass, suggesting that the energy effect is dependent on body composition and on fat cell thermogenesis [256]. A double PC1/3 mutant with Gln665Glu/Ser690Thr pair does not exhibit a significant loss of activity nor changes in secretion or maturation *in vitro*. However, it would be interesting to test whether these mutations affect prohormone processing in cells possessing a regulated secretory pathway. Interestingly, the Asn221Asp SNP always coexists with the Gln665Glu/Ser690Thr SNP cluster but not vice versa, suggesting that the Asn221Asp SNP exists on the minor allele. PC1/3 proteins containing the three SNPs Asn221Asp, Gln665Glu, and Ser690Thr exhibit enhanced PC1/3 maturation, although when transfected into GH4C1 cells, POMC processing is not affected [257].

Four other studies have found linkage between a region on chromosome 5q that contains the *Pcsk1* gene and obesity [258, 259, 260, 261]. Other PC1/3 SNPs that are also shown to be associated with obesity include the intronic rs155971 and rs3811951 in a Chinese population [262]. Finally, PC1/3 SNPs are associated with taller body height and wider chest circumference in goats [263], indicating that PC1/3 could potentially represent a marker for body size across species.

2.7.3   PC1/3 and Other Diseases

In a leptin-resistant mouse model, transplantation of PC1/3-expressing pancreatic cells increases pancreatic PC1/3 activity, successfully treating diabetes. This implies that targeting PC1/3 might be an effective therapeutic approach to treating diabetes or other endocrine disorders [264]. PC1/3 may also be an effective target in cancer. For example, silent corticotroph adenomas and prolactin adenomas exhibit decreased expression of PC1/3 [195]. In contrast, patients with nonfunctioning macroadenomas have higher serum anti-PC1/3 autoantibody levels compared to lymphocytic hypophysitis, other pituitary diseases, and healthy controls [265]. Changes in PC1/3 expression and activity have also been implicated in melanoma, in non-small cell lung cancer, and in colorectal liver metastases. When Blanchard *et al.*, chemically induced tumor growth in transgenic mice with targeted expression of PC1/3 in the mammary epithelium, enhanced growth of normal and neoplastic mammary tissue growth was observed [266, 267]. These data suggest that PC1/3 may also be utilized as a biomarker to diagnose certain types of cancer. Interestingly, a statistically significant association between *Pcsk1* polymorphisms and longevity in Koreans has been reported [268].

2.7.4   ProSAAS and Disease Relevance

ProSAAS immunoreactivity has been found in neurofibrillary tangles and neuritic plaques of brain tissues from patients with Pick's disease, Alzheimer's disease, and Parkinsonism–dementia complex, implying that proSAAS may be involved in the pathophysiology of general tauopathies [82, 83]. In our laboratory, we have found co-localization of proSAAS with $A\beta_{1-42}$, $A\beta_{1-40}$, and tau in a transgenic mouse model of Alzheimer's disease, as well as with α-synuclein in human Parkinson's disease brain tissue (A. Hoshino, M. Helwig *et al.*, in preparation). In addition, four independent proteomic studies have identified proSAAS as a candidate biomarker for neurodegenerative diseases, including Alzheimer's disease, frontotemporal dementia, and Parkinson's disease [84, 85, 269, 270]. At this stage, it is difficult to conclude whether proSAAS plays a role in disease development or progression; in any case, these studies point to the potential for using proSAAS as a novel therapeutic target and/or a biomarker for neurodegenerative disease.

.   .   .   .

CHAPTER 3

Prohormone Convertase 2

3.1 INTRODUCTION TO PROHORMONE CONVERTASE 2

3.1.1 Unique Features

Prohormone convertase 2 (PC2) is an exceptional convertase in many ways. It is the only convertase capable of exit from the ER with an uncleaved propeptide; it is the only convertase to require the assistance of a secretory pathway chaperone, 7B2 for manifestation of any enzymatic activity; and it is the only convertase to contain an Asp rather than an Asn in the oxyanion hole (reviewed in [183]). As will be described further below, these properties have all evolved to suit its particular need to perform cleavage reactions within the late secretory pathway.

3.1.2 Evolution

Peptidergic signaling is evolutionarily ancient; PC2 has been identified in 26 species to date (Gen-Bank), from insects [271] to worms [272, 273] to *Aplysia* [274], and therefore can be considered an evolutionarily old convertase. Interesting species in which the PC2 sequence has been identified include sea urchins, planaria, a variety of insects and spiders, and many different types of worms. While PC2 has not been found in hydra, which contains only three unusual PC1/3-like enzyme(s) (D.F. Steiner and S. Chan, *personal communication*), it has been identified in sponge (*Amphimedon queenslandica*) [275]. Evolutionary diagrams of convertase evolution have been published by Bertrand *et al.* [276], who studied convertases in *Amphioxus*, and by Srivastava *et al.*, who studied sponge convertases [275]. Interestingly, sponges appear to have five different PC2-like genes (as well as several other members of the convertase family), suggesting an increased need for peptidergic communication in early eukaryotic evolution. However, these sequences need to be verified further since, as discussed above for PC1/3, assignation of convertase identity to a particular subfamily is difficult in early eukaryotes. The PC2 binding protein 7B2 has been identified in *Aplysia, C. elegans,* insects, and in most vertebrate genomes; it has not yet been identified in sponge or hydra. This evolutionary spread is much larger than that of proSAAS, thus far only identified in vertebrates (and quite divergent even within different vertebrates), suggesting earlier evolution of 7B2.

3.1.3   Domain Structure and Functions

ProPC2, like PC1/3, contains four different domains (shown in Figure 7). These are the propeptide, which, as in other convertase propeptides, contains an internal (secondary) cleavage site (shown with arrow); the catalytic domain, which carries out catalysis; the P domain; and a short C-terminal tail domain. Each will be discussed in turn. The PC2 **propeptide**, like that of other convertases, is capable of inhibiting PC2 via the tetrabasic sequence at the primary cleavage site; however, it is a fairly weak inhibitor compared to the propeptide of other convertases [42]. Like the propeptide of subtilisin, the PC2 propeptide serves as an intramolecular chaperone, as shown by the fact that mutations introduced within the propeptide can affect the structure of the mature enzyme, even though the mature enzyme no longer contains the propeptide [42]. The propeptide is important for proper synthesis [50, 277, 278] and may also direct proPC2 to the regulated granule compartment via association with granule membranes [279]; propeptide-mediated membrane association of the enzyme has been shown [280]. Lastly, the PC2 propeptide also interacts with the PC2-binding protein 7B2 [281].

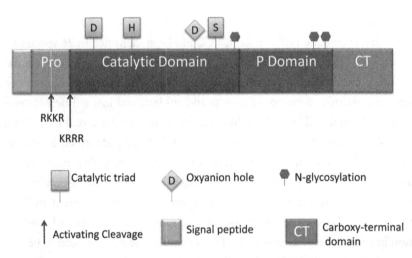

FIGURE 7: Domain structure of preproPC2. The primary cleavage site is marked by an arrow; the secondary cleavage site is marked with a white dotted line. The residues which accomplish catalysis, His, Asp and Ser, are shown with purple boxes; the oxyanion hole is occupied by an Asp (orange diamond). Putative sugar sites are indicated with pink hexagons. The P domain, also known as the "middle" or HomoB domain, lies adjacent to the catalytic domain and is required for enzyme expression. The carboxy-terminal domain is thought to contribute to efficient enzyme sorting.

The **catalytic domain** of PC2 is highly conserved with those of other PCs and contains the subtilase triad responsible for effecting catalysis; however, as mentioned above, in PC2, the characteristic Asn in the oxyanion hole which stabilizes the substrate transition state is substituted by Asp. This substitution may represent an accommodation to the low pH in which this particular eukaryotic subtilase operates; the pH of the mature secretory granule is thought to be around 5.0.

The **P domain**, consisting of an eight-stranded beta barrel [54, 59], is critical for both expression as well as zymogen cleavage [50] and contributes to both calcium binding and pH optimum [58]. Truncation past a critical threonine activity at the end of the last strand of the beta barrel results in total loss of activity and expression [58], most likely due to protein instability. Creemers *et al*. have shown that this domain promotes the efficiency of the sorting process [277].

Unlike the convertases furin and PC1/3, in which several domains can be interchanged without loss of expression and functionality [50, 277], PC2 generally requires its own domains for proper expression [50, 58, 277, 282], suggesting unique domain interactions. However, the **carboxyl terminal domain** of PC2 can act as a sorting domain when transferred to furin, implying that it is responsible for granule targeting; the P domain further enhances storage efficiency [277]. Deletion studies confirm the requirement of the C-terminus for sorting [283]; others have shown that the C-terminus of the enzyme plays a role in membrane association, which is likely relevant to sorting to secretory granules [284]. In general, convertase C-terminal domains are the least conserved segments between different convertases and between different species. While vertebrate C-terminal PC2 tails are very well-conserved, invertebrate tails show considerable divergence. Interestingly, the paired basic residues present in this domain are conserved in invertebrate PC2s, although they are not known to be cleaved. The PC2 tail is the shortest of all of the convertase tail domains.

Modeling of the two prohormone convertases has been performed based upon the known crystal structure of furin [285]. These models show differences in binding pocket architecture which presumably contribute to differential specificity [52] and also show that the P domain is likely to have a similar structure in all of the proprotein convertases. Two bound calciums are predicted in the general convertase structure [286]; it is possible that additional calcium binding sites exist. In this PC2 model, the smaller number of negative charges lining the non-prime subsites in the substrate binding pocket explains the reduced requirement for basic residues N-terminal to the substrate cleavage site as compared to furin [285].

3.1.4   The PC2-Binding Protein 7B2

The first demonstration that proPC2 has a specific binding protein was performed by Braks and Martens [287] who showed using coimmunoprecipitation that the small neuroendocrine protein 7B2 can bind to PC2, a finding then confirmed by two other groups [288, 289]. However,

Structure of Vertebrate 7B2

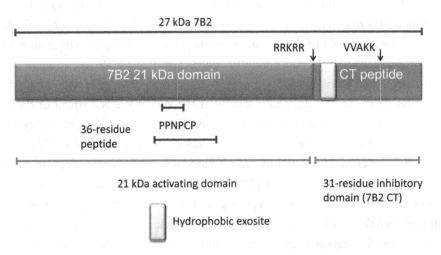

FIGURE 8: Domain structure of 7B2. The 185-residue 7B2 protein consists of two functional domains, a 21 kDa amino terminal domain (blue) which blocks proPC2 aggregation, and a 31-residue inhibitory peptide (green). The 36-residue peptide which contributes most to the anti-aggregant activity against proPC2 is underlined. All 7B2s discovered thus far, from invertebrate to human, contains the invariant PPNPCP motif within this peptide. The carboxy-terminal domain (green) is responsible for PC2 inhibition via its amino terminal segment terminating in VVAKK; removal of the KK residues by CPE is required to end inhibition.

the function of this binding event was not evident until later work showed that co-expression of 7B2 is absolutely required for acquisition of PC2 enzymatic activity [289, 290, 291]. The acidic, tyrosine-sulfated 7B2 protein was discovered in 1984 [292, 293] and named after a chromatographic fraction. 7B2 has been described as an excellent neuroendocrine marker [294], as it is found in every neuronal and endocrine tissue and almost every cell line derived from such tissues (SK-N-MC cells being the only known exception). 7B2 does not bind to other convertases [288].

The domain structure of 7B2 is shown in Figure 8. The 185-residue protein (which has a 186-residue isoform) consists of two domains separated by a furin cleavage site [295]. While the amino-terminal 21 kDa domain is responsible for the facilitation of proPC2 maturation [289], the carboxy-terminal 31 residues represent a potent inhibitor of PC2 [296, 297, 298] inactivated at the Val–Val–Ala–Lys–Lys site via slow cleavage by PC2 [299]. All known 7B2s contain the signature sequence Pro–Pro–Asn–Pro–Cys–Pro, but are otherwise little conserved, with the exception of the inhibitory CT peptide which always contains a sequence similar to Val–Val–Ala–Lys–Lys. 7B2 sequences have been identified in worms [272], flies [300], and in many other species, though not yet in early eukaryotes such as hydra and sponge (even though these species are thought to contain PC2). The inhibitory sequences are often duplicated in lower eukaryotes.

3.1.5   Gene Location

The PC2 gene, named *Pcsk2*, is located on chromosome 20 (p11.2) in humans and contains 12 exons [5, 87]. The 7B2 gene (also known as *Scg5* and *Sgne1*) is on chromosome 15 (q13) and has six exons [301]. In mice, the PC2 gene is located on chromosome 2; the 7B2 gene is located about 30 centimorgans away on the same chromosome [301], in an area generally rich in genes related to obesity [302].

3.2   DISTRIBUTION

3.2.1   Tissue Distribution

Early studies showed that PC2 is expressed predominantly in cells maintaining a regulated secretory pathway, such as brain and peripheral nervous tissue, and endocrine cells, such as pancreas, where it appears predominantly in alpha cells; adrenal medulla; and pituitary [91, 93, 94, 95, 96]. In the rodent pituitary, PC2 is exclusively expressed in the intermediate lobe [93], and in the human pituitary, in areas rich in α-MSH [101]. An early *in situ* hybridization study showed that PC2 message is rich in rat hippocampus, habenula, and various thalamic and hypothalamic nuclei [199]. PC2 has been identified in the superior cervical ganglion, where it acts to process proNPY [303]. PC2 has also been identified in the retinal tissue of the eye [100, 304], thyroid C cells [105, 106], and in the immune system [116, 117, 305]; very low amounts are also found in the skin [306]. Beaubien and colleagues have demonstrated the presence of this enzyme within aortic ganglia [307].

3.2.2   Intracellular Distribution

Within the neuroendocrine cell, the bulk of proPC2 appears to reside in the ER [67, 145, 308, 309]; mature PC2, however, is predominantly localized within regulated secretory granules [68, 308]. This distribution is unusual among convertases, though PC7 seems to exhibit similar storage in the ER [310]. Binding of PC2 to various types of membranes has been demonstrated by several groups [154, 279, 280, 284]. Differential localization of immunoreactivity for PC1/3 and PC2 in granules from the rat anterior pituitary was shown by electron microscopy [311]; PC2 was described as being in both large lucent granules (together with secretogranin II) and small dense granules, but PC1/3 was only observed in large lucent granules.

3.2.3 Cell Lines

PC2 expression is high in two pancreatic cell lines, the rat insulinoma Rin5f and the mouse insulinoma β-TC3 and the pituitary lines GH4C1 and GH3 [92, 290]; however, Waldbieser *et al.* [312] found no expression in GH3 cells. The neuronal lines NG108 and Neuro2a contain small amounts of PC2 [91]. The adrenal cell line PC12 is also rich in 7B2, but contains no PC2 ([312]; M. Helwig and I. Lindberg, unpublished results). Host cell lines lacking PC2 expression, such as AtT-20, PC12, and GH4C1, have been useful to study the effects of transfected enzyme and substrate. Similarly, the SK-N-MCIXC neuroepithelioma cell line has been a useful model to test the effect of 7B2 transfection on proPC2 cleavage and activation, as this line expresses proPC2, but no 7B2 [133, 290]. βTC-3 stable cell lines expressing the chimeric tRNA^{Val}-δ ribozyme transcript targeting PC2 mRNA have been used to establish the contribution of PC2 to specific processing events [313]. The stable AtT-20/PC2 cell created by the Mains group [129] represents an especially valuable resource to characterize the contribution of PC2 to endocrine processing; for example, augurin processing by PC2 was ruled out by this mechanism [314]. Rat bone marrow stromal stem cells can be induced to express PC2 upon neural differentiation [133]; expression of 7B2 was not examined in this study.

3.2.4 Development

Marcinkiewicz [137] demonstrated that PC2 transcripts are detectable in the presumptive adenohypophysis beginning on embryonic day 15 (E15). In the intermediate lobe, PC2 mRNA first appears between E16 and E18 and its level increases during development, reaching maximal expression in the adult. PC2 mRNA expression peaks between postnatal days 1 (P1) and 14 (P14) and then decreases to adult levels. In the anterior lobe, during the P1–P14 postnatal period, PC2 immunoreactivity can be detected within cells expressing α-MSH immunoreactive peptides [137].

In studies of mouse pancreatic development, the first detectable expression of PC2 was observed in the pancreatic primordium at midgestation on embryonic day 10. Immunocytochemical experiments confirmed the presence of immunoreactive PC2 on embryonic days 14 and 17 [315]. Holling *et al.* [316] have described the development of PC2 and 7B2 message in the frog; surprisingly, 7B2 message precedes PC2 message and is apparent in the oocyte.

3.3 CELL BIOLOGY AND MATURATION
3.3.1 ProPC2 Maturation

Figure 9 depicts the maturation of proPC2 to its active form, in association with its binding partner 7B2. ProPC2 is initially synthesized as an N-glycosylated proform [308] which is later tyrosine-sulfated [281] and sialylated; however, not all of its sugar chains can mature to sialylated forms

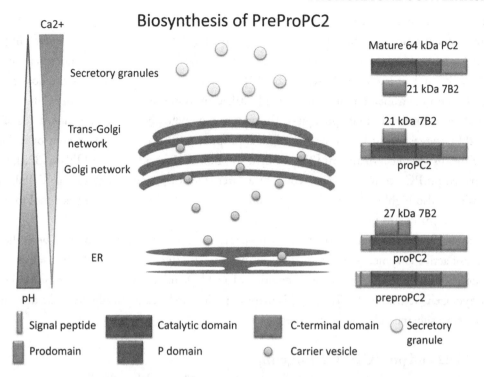

FIGURE 9: Biosynthesis of PC2 and 7B2. This slide shows the progressive maturation of proPC2 through the secretory compartment. 7B2 and proPC2 bind each other in the endoplasmic reticulum; in the trans-Golgi network, the furin site within 7B2 is cleaved. The complex then gets packaged into immature secretory granules, where progressive acidification results in autocatalytic activation of proPC2. 7B2 is present throughout the secretory pathway, most likely to block unproductive zymogen aggregation, which results in an inactivatable precursor.

since mature PC2 contains both endoglycosidase H-resistant and -sensitive sugars [67, 145, 277, 308, 309]. We have found that deletion of sugar sites by mutagenesis results in loss of its ability to exit the ER, suggesting that glycosylation is required for stability (J. Hwang and I. Lindberg, unpublished observations).

An intermediate 71 kDa precursor form contains the C-terminal portion of the PC2 propeptide; this cleavage does not seem to be obligatory for primary site cleavage [42]. This site is not cleaved in CHO cells, but is cleaved in neuroendocrine cells; cleavage is blocked by bafilomycin, suggesting possible involvement of PC1/3 [317]. Many studies have shown significant ER retention of the zymogen compared to proinsulin [308], POMC [67, 145], and 7B2 [318]. Like all subtilases, proPC2 undergoes autocatalytic activation to a smaller form; the PC2 zymogen is catalytically inactive. Activation is an intramolecular autocatalytic process which occurs spontaneously for puri-fied recombinant proPC2 when the pH is lowered to 5.0 [319]; it is not calcium-dependent [183]. A proPC2 mutant in which the catalytic Ser has been mutated to Ala does not undergo propeptide removal, although it is efficiently secreted [317]. Although one report suggests the possibility of partial C-terminal truncation of mature PC2 [320], antiserum directed towards the C-terminus of proPC2 can be used to immunoprecipitate the mature enzyme without time-dependent loss of label [139, 289], suggesting little if any truncation. Disruption of the acidic environment of endocrine cells, either by chemical treatment [321] or by disruption of the acidifying v-ATPase [322], results in impaired proPC2 conversion and reduces the later steps of precursor processing. Cellular PC2 processing is also highly sensitive to temperature [69, 321] and this can be used as a tool to distin-guish PC2 from PC1/3 processing [128].

Interestingly, the removal of the PC2 propeptide is not necessarily associated with the pro-duction of active enzyme, as "unproductive" propeptide cleavage, resulting only in inactive mature enzyme, is common [317]. Thus, the observation of a 64 kDa mature species does not *per se* demon-strate zymogen "activation" [323] since this mature species can be catalytically inert; the demonstra-tion of active substrate cleavage is required.

3.3.2   7B2 and proPC2/PC2 Targeting

The small neuroendocrine protein 7B2, discovered in 1984 by Seidah and colleagues as a neuroen-docrine marker [293] is now known to act both as a potent inhibitor of active PC2 [296, 297] as well as to paradoxically be required for the production of active enzyme, a process we have termed "capacitation" [324]. Unlike PC1/3, the expression of proPC2 in constitutive cells does not result in the release of active enzyme [309]; 7B2 must be co-expressed for proPC2 to retain its capacity for later activation [289]. 7B2 does not perform the function of a classical chaperone, for example, by assisting initial folding of proPC2 in the ER; indeed, proPC2 folds independently of 7B2 and cannot bind 7B2 until it has finished folding, a process that is quite slow [318]. Once proPC2 has folded and has bound 7B2, the complex exits the ER (note that 7B2 binding is apparently not oblig-atory for exit, as proPC2-transfected CHO cells lacking 7B2 expression secrete abundant—though inactivatable—proPC2 [309]). Unlike all other convertases, proPC2 can be secreted in the absence of propeptide cleavage. However, unless 7B2 is present during proPC2 synthesis [289] or is pres-ent in the secretion medium [324], any resulting smaller-sized enzyme will be totally enzymatically inactive.

Activation of proPC2 to PC2 does not appear to require the continued presence of 7B2 [319, 325]; although cell labeling studies indicate that 7B2 may still be associated with the protein following activation [288, 325], the affinity of 7B2 for PC2 is clearly weakened upon propeptide cleavage [325]. In certain tissues such as islets and pancreatic cell lines, 7B2 can be phosphorylated, a process which results in its inactivation; phosphorylation does not occur in AtT-20 cells [326]. Extensive structure–function studies support the idea that a well-conserved 36-residue disulfide-bonded peptide within 7B2, residues Gly86-Asp121, can assume much of the capacitation function [327, 328]; this peptide is predicted to contain polyproline II and alpha helices [327], and mutations which disrupt these secondary structures abrogate binding [327, 328]. However, another structure-function study pointed to a more amino-terminal segment, His58-Pro88, and especially of Tyr87, as important for PC2 function [298]. A third study implicated the furin site as a binding determinant for PC2 [288], at odds with the observation that the amino-terminal domain of 7B2, which lacks this region (Figure 8) is sufficient to rescue PC2 activity [289] and other data showing that a mutant 7B2 lacking a cleavable furin site can both bind proPC2 and facilitate its maturation [329]. However, in this latter case, the resultant enzyme remains inhibited, suggesting that the inhibitory sequence within the 7B2 CT peptide is unable to exit the PC2 active site [329]. Cleavage at the furin site is apparently required to escape inhibition, perhaps by relieving conformational restrictions [329]. Obtaining structural information for the various 7B2–proPC2 complexes would shed much light on the complex interaction of these two proteins.

Recent experiments show that the major function of 7B2 is not to directly facilitate proPC2 maturation, but instead to block the spontaneous aggregation of proPC2 into oligomers and aggregates which are enzymatically incompetent for activation, i.e., do not yield active enzyme [324]. Thus, 7B2 acts as an anti-aggregant chaperone, an interesting process which it may also perform for other secretory proteins prone to aggregation [330]; M. Helwig and A. Hoshino *et al.*, unpublished results).

Several studies have attempted to identify binding determinants for 7B2 in the proPC2 structure. There is a clear requirement for a Tyr194-containing hydrophobic loop and the PC2 propeptide for tight binding of 7B2 [281, 282, 331]; the oxyanion hole Asp is also involved in binding 7B2 [331]. Further work has shown that the PC2-specific residues 242–248 (the P4 canopy region) are involved both in propeptide maturation and in inhibition by the 7B2 CT peptide [332]. Interestingly, when this sequence is swapped into the corresponding area of PC1/3, proPC1/3 is well expressed, but its propeptide fails to undergo cleavage [176], and it remains trapped in the ER.

3.4   ENZYMATIC CHARACTERIZATION

3.4.1   General Enzymatic Properties

Like PC1/3, PC2 is a calcium-dependent enzyme maximally active at low pH; its pH optimum of 5.0 is lower than that of PC1/3. The concentration of calcium required for enzymatic activity is also lower than that of PC1/3, about 100 μM [3, 166, 320]. The substrate turnover rate of PC2 is more similar to that of furin than that of PC1/3, about 30 μmol amc/mg/h. While others have termed this enzyme "intrinsically crippled" [333], as described in Section 2 above, the turnover rate of PC2 is actually 80–100× higher than that of PC1/3 (which therefore by comparison might be thought of as catalytically moribund). Indeed, while mammalian subtilase turnover rates are apparently quite low compared to digestive, blood and lysosomal proteases, it should be noted that prohormone convertases and substrates may be unusually concentrated within the secretory pathway (perhaps both are even membrane-associated), and also spend lengthy periods of time together inside secretory granules. K_m values of PC2 for fluorogenic substrates vary from 18 to 131 µM (see [183]) and are not improved using natural peptide substrates. Interestingly, swapping the uncharacteristic Asp in the oxyanion hole, which stabilizes the transition state, for the typical Asn found in other subtilases, does not increase the catalytic activity of PC2 [50, 334]. Rockwell *et al.* [333] have shown that peptidyl methyl coumarin esters can be used to perform active site titration of PC2, due to slow deacylation rates.

It is quite difficult to purify active PC2 from natural sources. A partially purified PC2 preparation from insulinoma granules was initially described by Hutton and colleagues [10]. PC2 has also been partially purified from bovine adrenal chromaffin granules [335] and as a recombinant protein, via baculoviral expression [336]; however, the high pH optimum found in these latter two studies does not correspond to that expected for a granule enzyme nor to that found in any other studies. DES cells (a *Drosophila*-based expression system) can be used to generate active recombinant PC2 [300], indicating that the use of insect cells is not a problem in this regard. Recombinant PC2 is however easily obtained as a zymogen from the conditioned medium of CHO cells engineered to express both this protein as well as 7B2; activity is associated with the zymogen form rather than the cleaved forms also present in the medium [166]. In the absence of 7B2 expression, proPC2 is well expressed and well secreted from CHO cells, but consists of only aggregated, inactive forms [309]. Addition of the 21 kDa domain to recombinant activated PC2 stabilizes the enzyme and protects it from heat denaturation [166]; substrate cleavage is also enhanced [337].

3.4.2 Contribution of Various Domains to Enzymatic Activity

The P domain has been shown to be involved in calcium recognition and pH optimum [58]. Carboxyl-terminal domain tethering of PC2 by addition of the PAM transmembrane and cytoplasmic domains affects cellular routing and increases enzyme activity [179], suggesting inhibitory action of the PC2 carboxy-terminal tail reminiscent of that seen with PC1/3 (see above). This latter finding explains the fact that it was possible to use C-terminally directed antiserum for immunoprecipitation of active PC2 from conditioned medium [338]. Indeed, this enzyme preparation yielded the only source of active PC2 until the discovery that coexpression of 7B2 was absolutely required to generate active PC2 [289].

3.4.3 Tissue Analysis of PC2 Activity

The ability of PC2 to undergo specific inhibition by the 7B2 carboxy-terminal peptide provided the basis for the construction of an assay for this enzyme in tissue and cell extracts [339, 340]. In this assay, it is critical to adequately inhibit the activity of the much more active proteolytic enzymes which can cleave the pERTKR-aminomethylcoumarin substrate, using a cocktail of enzyme inhibitors. The specific PC2 activity is given as the difference between that measured in the presence and absence of the 7B2 CT peptide, a specific PC2 inhibitor (see inhibitor section below), and represents from 50% to 90% of total enzymatic activity, depending on the tissue. Calcium and detergent are required in the assay, as well as the continued presence of nonspecific protease inhibitors. The use of a PC2-specific inhibitor provides confidence that the enzymatic activity being measured can actually be attributed to PC2.

3.4.4 Substrate Specificity

PC2 possesses broader substrate recognition than PC1/3, possibly indicating a larger substrate binding pocket. As one of many examples, PC2 easily cleaves the model substrate ACTH internally to generate ACTH 1–17; PC1/3 is incapable of this reaction. This cleavage also exemplifies the general observation that PC2 often performs the final endoproteolytic steps required for peptide hormone maturation, for example, the cleavage of small enkephalins from enkephalin-containing intermediates [162, 167], or the cleavage of α-MSH from the precursor ACTH [95, 148, 341]. A list of prohormone/neuropeptide substrate cleavages in which PC2 action is implicated is shown in Table 2. (Many of the cleavage events listed in this table are also documented in the Sanford–Burnham's index of proteolytic cleavages (http://cutdb.burnham.org/loginCutDb).) This table includes both transfection analyses, in which cDNAs for peptide precursors are transfected into PC2-expressing and non-PC2 expressing cell lines; the results of *in vitro* incubations with PC2; and the results of proteomic studies using convertase knockout mice [185, 186]. In the latter study, extracts are prepared from brains from animals in which PC2 is not expressed; thus sites normally cleaved by PC2 remain intact, permitting identification of PC2-specific sites. Substrate specificity has also been explored using a series of synthetic substrate analogs, as well as by using internally quenched peptides based upon the known substrate proenkephalin [167]. Peptide substrates based on a series of viral and bacterial proteins were also examined in a general convertase specificity study [342], though these particular viral and bacterial sequences are biased for furin cleavage and as such are unlikely to encounter PC2 in a cellular context.

Cleavage preferences obtained from these studies are summarized in Figure 10 (see also details in [183]). While uncommon, substrates containing certain large residues (Tyr, Trp, and Pro) in the P1' and P2' positions are preferentially cleaved by PC2 rather than by PC1/3, though both

PC2 Specificity

P6	P5	P4	P3	P2	P1	P1'	P2'	P3'
	R,D,L,E	R,E,G,L	Q,E	K	R	S,G,A	S,G,A	G,S,N,L
P,G,L,E	L,E,R,S	R,L,F,V	E,Q,L	K,R	R	A,S,G	G,S,L	E,G,Q

Note: although not preferred, peptides
containing Y, W, or P at P1' and P2' are cleaved
almost exclusively by PC2

FIGURE 10: Specificity of PC2 for various residues in the enzyme pocket. The data in the top row represent the top-scoring residues (single letter amino acid code) and originate from the proteomics studies of the Fricker laboratory [185, 186]. The data in the second row originate from all sources: literature (see Table 2), unpublished work, and proteomics studies. We thank A. Ozawa for compiling these data.

enzymes prefer small neutral residues such as Ser and Ala at this position. A substrate containing Pro in the P1' position can be cleaved by PC2, but is not favored, since substitution by Ala improves efficiency [343]. The P1 position must be occupied by a basic residue; here, Arg is preferred over Lys. The P2 position must also be occupied by a basic residue, though PC2 can cleave dynorphin A1–17 at a single basic residue, possibly because of the presence of additional basic residues at P3 and P4 [343]. Lys seems to be preferred over Arg at the P2 position [342]. The P3 and P5 positions do not seem to contribute greatly to specificity, though proteomics analysis indicates some preference for a charged residue at these subsites [185, 186]. An Arg at the P4 position may confer slight benefit [167, 342] although most prohormone substrates do not contain P4 basic residues. PC2 does not appear to prefer longer substrates since lengthening did not affect K_{cat}/K_m [167]. PC2 seems to be the only enzyme that can cleave the Lys–Lys bond of POMC [67], although a Lys–Lys sequence can be cleaved by PC1/3 when present in proenkephalin in these same cells [344]. Interestingly, very few known precursors are totally resistant to PC2 cleavage; one example is proghrelin, cleaved only by PC1/3 and furin, but not by PC2 [177, 445].

It should be noted that neither cleavage predictions made with online programs such as Neuropred (http://neuroproteomics.scs.illinois.edu/cgi-bin/testneuropred.py) nor the results of actual *in vitro* incubations of recombinant enzyme with recombinant substrate are necessarily physiologically relevant, since in cells, the ability of PC2 to cleave a given precursor depends on the cellular itinerary of the precursor as well as the enzyme. In other words, a constitutively secreted precursor may have no opportunity to encounter PC2 in regulated secretory granules and will thus not be cleaved by this enzyme in neuroendocrine tissues.

In summary, while much information has been gained during the last two decades on the specific amino acid requirements for prohormone susceptibility to PC2 cleavage, a full understanding of substrate specificity will likely require crystallographic mapping of the binding pocket.

We have used mutagenesis to examine the contribution of PC2-specific residues to its relatively broad specificity; corresponding residues from PC1/3 were placed into the substrate binding pocket of PC2 [345]. Interestingly, a residue swap in the S6 position produced the largest changes in specificity, with the enzyme acquiring some PC1/3-like characteristics. However, overall, we were not able to convert PC2-like specificity to PC1/3 activity, suggesting that structural information outside the substrate pocket must contribute significantly to specificity [345].

3.5   REGULATION OF EXPRESSION AND ACTIVITY

3.5.1   Transcriptional and Translational Regulation

Unlike PC1/3, PC2 expression is not regulated on a translational level by glucose [213]; however, transcriptional regulation of PC2 by glucose has been reported [107]. Thyroid hormones also stimulate PC2 synthesis via increased transcription [189, 346, 202]; the PC2 promoter contains thyroid response elements [189]. Early studies noted the presence of AP-1 and Sp1 binding sites, cAMP-responsive elements (CRE); an interferon consensus sequence (ICS); and three POU proteins (e.g., GHF-1) binding elements in the mouse PC2 promoter [88]. Evidence has been presented as to the involvement of a CRE sequence in drug addiction [202]. Others have shown that different cell lines are modulated differentially by second messenger systems in different cell lines [197]; for example, phorbol esters increase PC2 message in WE4/2 cells, but not in SK-N-MCIXC cells. The transcription factor Pax6, important in pancreatic development, regulates the transcription of both PC2 and 7B2 indirectly through cMaf and Beta2/NeuroD1 [347]; the transcription factor Egr-1 has been shown to modulate PC2 expression [348]. A binding site for the repressor RE1/NRSE has also been identified within the PC2 promoter [349], providing a potential explanation of why PC12 cells—which greatly overexpress this repressor [350]—do not express PC2, while the parental tissue, adrenal medulla, does express this protein. Lastly, the neuronal transcription factor nescient helix–loop–helix 2 protein (Nhlh2) regulates PC2 transcription [351].

Dopaminergic agonists such as bromocriptine suppress pituitary message for PC2 in the intermediate pituitary, while haloperidol, a dopamine antagonist, has the reverse action, increasing PC2 message [352, 129, 208]. Indeed, the D2 receptor knockout mouse exhibits a 3- to 4-fold increase in pituitary PC2 expression [353]. Electroconvulsive shock has also been shown to raise

PC2 message in the hippocampus [354]. These results indicate that PC2 expression can undergo dynamic changes depending on the physiological state of the animal.

3.5.2 Endogenous Inhibitors

The C-terminal 31 residues of 7B2, the 7B2 CT peptide, represents a potent inhibitor of PC2 *in vitro* [296, 355]; while intact 7B2 is a competitive inhibitor, the rates of dissociation are so slow that its kinetics resemble those of noncompetitive inhibitors [296]. However, when directed to secretory granules via fusion with a peptide hormone, the 7B2 CT peptide does not inhibit precursor cleavage [356], indicating that the major function of this peptide may not be to control peptide cleavage within the secretory granule, but rather to serve another function—possibly to block premature proPC2 cleavage in the early stages of the secretory pathway. It is interesting to note that a full 16 residues of this peptide are required for potent inhibition of PC2 [297]; a hydrophobic exosite in this peptide is conserved down to invertebrates [183] and clearly plays an important role in tight binding of this inhibitor to the enzyme. The 7B2 CT peptide is inactivated by cleavage at the VVAKK site; this must be followed by CPE removal of C-terminal basic residues, as a peptide terminating in VVAKK is still highly inhibitory [299].

Recent data indicate that catecholamines can inhibit PC2 activity [170]. These inhibitory concentrations are much lower than those known to occur within chromaffin granules, suggesting that catecholaminergic inhibition may contribute to control of PC2 activity in adrenal chromaffin cells. Whether inhibition also occurs in other catecholaminergic tissues, for example brain, remains to be determined. An intriguing study from 1992 in which a bovine chromaffin granule extract was probed with antiserum to glycoprotein 3 (clusterin) showed tight association of PC2, PC1/3, and CPE with this chaperone protein [357]. Whether clusterin can affect the activity of these enzymes has not yet been explored; interestingly, a clusterin-related protein, SPARC, associates with PC1/3 during expression in CHO cells (I. Lindberg, unpublished results).

CPE was shown to activate PC2 cleavage of prodynorphin 1–17 [22], suggesting potential feedback inhibition by basic-residue extended peptides for this substrate. Further work is necessary to establish whether cleavage of other substrates is enhanced by the presence of CPE.

Propeptides represent powerful endogenous inhibitors of furin and PC1/3 [358]; however, the propeptide of PC2 is a comparatively weak inhibitor of this enzyme [42], perhaps because the 7B2 has evolved as a separate mechanism to suppress premature activation of this enzyme. A synthetic peptide corresponding to the 28 residues preceding the catalytic domain is a micromolar inhibitor of PC2 [42]. The role of internal cleavage of the propeptide is also unclear; while this is clearly rate-limiting for the formation of active furin [359], internal cleavage does not seem to be regulatory for PC2 cleavage [42].

The testicular protein CRES (cystatin-related epididymal protein) has been shown to potently inhibit PC2 in a competitive fashion [360]. CRES was identified in the pituitary as well as in β-TC3 cells, suggesting possible physiological relevance. This inhibitor is inactive against cysteine proteinases; the mechanism by which it could inhibit PC2 is unclear. Lastly, we have found that micromolar concentrations of ascorbate inhibit recombinant PC2 (M. Kacprzak and I. Lindberg, unpublished observations). This effect appears to be related to the reducing capacity of this molecule since oxidized ascorbate is ineffective. How PC2 is able to escape inhibition by the high concentrations of ascorbate present in chromaffin granules is not clear; addition of ascorbate to chromaffin cell cultures does not affect PC2-mediated prohormone processing (M. Kacprzak and I. Lindberg, unpublished observations).

3.5.3   Synthetic Inhibitors and Activators

Few studies have addressed the pharmacologic inhibition of PC2. While the long length required for inhibition by the 7B2 CT peptide [297] would seem to indicate that small molecule inhibitors will not be identifiable, and indeed combinatorial peptide library screening failed to identify six-residue inhibitors (I. Lindberg, unpublished observations), recent studies suggest that it will be possible to synthesize potent small molecule inhibitors of PC2. Using combinatorial compound library screening, we have identified two scaffolds, bicyclic guanidines and pyrrolidine bis-piperazines, that when appropriately modified, resulted in micromolar inhibitors of PC2 [169]. The best inhibitor was a pyrrolidine bis-piperazine with a Ki of 0.5 μM and great specificity (Kis greater than 25 μM for inhibition of furin and PC1/3). More recent work has identified modified 2,5 dideoxystreptamines that are micromolar inhibitors of PC2 (M. Vivoli et al., in press). Excitingly, activators of PC2 were also identified in this work, and modeling predicts the presence of allosteric sites (M. Vivoli et al., in press). While no endogenous activators have been found as yet, if the presence of allosteric sites is validated by further work, it will be interesting to search for such molecules.

3.5.4   Regulation of PC2 Activity by its Chaperone 7B2

Several papers have provided evidence for coordinated expression of 7B2 and PC2; for example, in *Xenopus* pituitary, PC2 and 7B2 are coordinately upregulated by dark/light adaptation [361]. Dexamethasone increases the expression of both PC2 and 7B2 in rMTC6-23 cells [362] and in pituitary cells [363]; in P19 cells, differentiation also results in coordinate upregulation of both proteins [364, 365]. However, given the requirement of proPC2 for 7B2 for the production of active enzyme, independent regulation of 7B2 would be an effective means to control PC2 activity. While early studies were disappointing in that they failed to show that overexpression of 7B2 exerted any effect on PC2-mediated POMC processing [281, 289], newly emerging evidence suggests that for at least one substrate, proglucagon, PC2 activity is indeed regulated by 7B2 expression [366, 367]. These studies show that overexpression of 7B2 is strongly correlated with increased glucagon levels and decreased 7B2 with lowered glucagon levels. This has now been demonstrated both in cell culture [367] as well as in two different animal model systems [368, 366], and appears to be specific for proglucagon since α-MSH production from POMC is unaffected. The molecular basis underlying 7B2-mediated substrate specificity still requires clarification, but may have to do with PC2 preference for certain precursors and peptide intermediates vs. others; slowly cleaved substrates may benefit more from additional PC2 activity, whereas for certain precursors, PC2 activity may not be rate-limiting. Fibroblast growth factor 23 appears to represent another substrate profoundly regulated by 7B2 expression [369]. It will be exciting in future work to determine the molecular basis for the substrate specificity of 7B2 regulation.

3.5.5   Regulation of 7B2 Expression

Increasing evidence suggests that 7B2 expression can be controlled by sequences within both the promoter as well as within an extended 5' regulatory region. The human 7B2 gene promoter contains a CRE, an AP-1 site, and several Pit-1/GHF-1-binding domains and heat shock element-like sequences [312, 370]. Interestingly, heat shock was able to increase biosynthesis and processing of 7B2 in *Xenopus* intermediate lobe cells, but did not increase message levels, suggesting an indirect mechanism of action, possibly translational control [370]. Medrano and colleagues have shown that inbred mice containing a specific sequence ("CAST") express greater amounts of 7B2 than mice lacking this sequence; overexpression of 7B2 is associated with decreased body mass [368]. Other workers have examined strain-specific differences in 7B2 expression among inbred mouse strains and have shown that expression of higher amounts of pancreatic 7B2 is associated with increased processing of proglucagon. All of these expression differences also appear to be due to promoter polymorphisms [366, 368].

New data suggest that the Beta2/NeuroD1, the proximal Pax6, and the Maf transcription factor binding sites may also be functional in the 7B2 promoter [347]; however, the Beta2/NeuroD1 site is not conserved in humans. Stat5b binding sites were observed in the study of Farber *et al.* [368]. Interestingly, the 5' untranslated region of the 7B2 message appears to severely restrain translation, potentially explaining the relatively low levels of the protein in various neuroendocrine tissues [371]. Recent work shows that 7B2 expression is also regulated epigenetically. Two papers by Waha and colleagues indicate that increased methylation of this gene is associated with reduced expression and increased tendency for oncogenic processes in medullablastoma and glioma [372, 373]. Given the realization that 7B2 levels can dynamically control PC2-mediated processing, more attention should be paid in future studies to determining the major mechanisms governing 7B2 expression in various physiological contexts.

3.6 MODEL SYSTEMS, KNOCKOUTS, AND MUTANTS

3.6.1 The PC2 Knockout

The PC2 gene was knocked out in 1997 by Steiner and colleagues; given that neuropeptide and hormonal processing is extensively disrupted, the null animals are surprisingly healthy (see Figure 11) [374, 375, 376]. These animals are however slightly runted and exhibit chronic hypoglycemia; this is likely due to their total lack of glucagon, since glucagon synthesis is abrogated [377]. Replacement of glucagon using osmotic mini-pumps restored glucose homeostasis [378]. These mice show hyperplasia of pancreatic delta and alpha cells, and prosomatostatin processing and proinsulin processing are also disrupted [375, 376]. Recent studies show that PC2 nulls exhibit reduced nociception in models of stress-induced analgesia, possibly due to lowered levels of endogenous opioids [379].

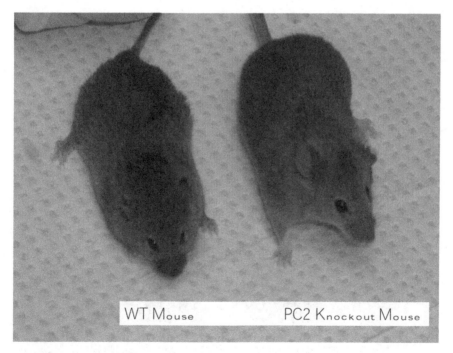

FIGURE 11: PC2 knockout mouse. Control and PC2 knockout mice are shown here; these mice are on the C57Bl6/J background and were obtained from the Steiner laboratory.

3.6.2 The 7B2 Knockout

7B2 knockout mice were produced by a retroposon-mediated strategy [291] and shown to exhibit a severe endocrine phenotype that consisted of small weight, hair loss, and death by 5–8 weeks (different colonies exhibit different longevity, with death in the Leder and Pintar labs occurring within 8 weeks, but within only 5 weeks in the Lindberg lab). A picture of the 7B2 knockout mouse is shown in Figure 12; these knockout mice, though often runted, appear quite normal until a few days before death. This early lethality was found to be due to intermediate lobe ACTH hypersecretion and ensuing hypercorticosteronemia coupled with a loss of glucagon [291]. This leads to extremely low blood sugar, the likely proximate cause of death [380]. Mortality can however be rescued by adrenalectomy, which eliminates the source of steroid [381]. The rampant corticosteronemia in the 7B2 null appears to be related to the low levels of pituitary dopamine in this knockout mouse [381]; dopamine normally plays an inhibitory role in intermediate pituitary hormone release. Aberrations in prohormone processing are directly due to loss of PC2 activity since PC2 is responsible for cleavage of ACTH to the α-MSH precursor ACTH 1–17 [148] and is the principal enzyme which generates glucagon from proglucagon [382].

7B2 Knockout Mouse   WT Mouse

FIGURE 12: 7B2 knockout mouse. The Leder laboratory 7B2 knockout mouse is shown on the left side. Note the shortened snout, thin tail and ruffled fur.

It should be noted that early work [291] indicated that the 7B2 null phenotype is independent of background strain. This is now known not to be the case. Knockout of 7B2 expression in the C57Bl6 mouse is not lethal; only in the background of a particular 129Sv substrain does the 7B2 knockout manifest its hypercorticosteronemic phenotype [383]. In this particular substrain, the PC2 and 7B2 nulls phenocopy each other [384]; thus the early assertion [291] that the major hormonal effects of 7B2 deletion differ radically from those of PC2 deletion is incorrect. These results highlight the need to use exactly the same substrain when comparing gene knockouts; for example, certain 129Sv strains are much more susceptible to glucocorticoid effects than are other mouse strains [385]. The commercially available (Jackson Laboratories) 7B2 knockout does not exhibit the lethal hypercorticosteronemic phenotype, which is now available only from the Pintar lab (UMDNJ). Like the PC2 knockout, the 7B2 knockout exhibits deficiencies in the processing of many neuropeptide precursors, for example CCK and gastrin [386, 387]. It is interesting to note that the original PC2 and 7B2 nulls process POMC somewhat differently, most likely due to strain differences in pituitary physiology [383].

Double knockouts for PC2 and 7B2 have also been created (Figure 13) (I. Lindberg and V. Laurent, unpublished data). These mice represent a cross not only of the two gene deletions but also of two different strains, C57Bl6/J and 129Sv. The resultant double knockout, observed in two lineages, showed a highly peculiar behavioral phenotype of inappropriate freezing behavior as well as extreme obesity (Figure 14). However, the phenotype was not stable and the entire strain has since been lost.

3.6.3   Model Systems

Several model systems in lower organisms have been used to study PC2 and 7B2 function. The *Drosophila* PC2 gene is known as *amontillado*, for its deleterious effects on hatching (the name originates from the Poe story in which a person becomes entombed). *Drosophila* PC2 has been shown to participate in embryogenesis and in larval and pupal development [271, 388], most likely through its participation in neuropeptide synthetic events [389, 390, 391]. Curiously, the 7B2–PC2 interaction seems to differ profoundly between insect and mammalian cells since mammalian proPC2 cannot mature to an active enzyme within constitutive cells (which lack a low pH storage compartment), but *Drosophila* proPC2 can do so [300].

C. elegans homologs of PC2 and 7B2 have also both been cloned. As in mammals, the *C. elegans* chaperone protein e7B2 assists in proPC2 activation [272]; later studies—in which e7B2 was renamed *sbt-1*—showed its involvement in synapse structure and function [392]. e7B2 was also found to affect worm lifespan [393]. *C. elegans* PC2 (CELPC2 [394], later renamed *egl-3* [395]), generates peptides which participate in mechanosensory responses [395]. Recent proteomics studies on the e7B2 and CELPC2 *C. elegans* knockout worms have demonstrated the involvement of these

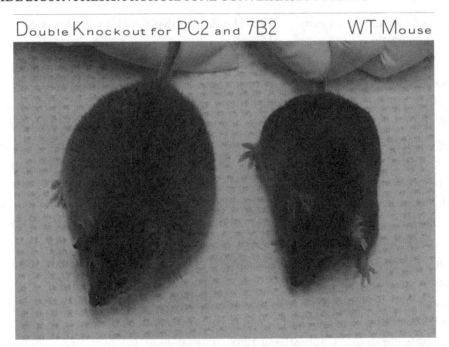

FIGURE 13: PC2-7B2 double knockout mouse. Mice obtained from the Steiner and Leder laboratories were bred together to form a double heterozygote strain. These were then mated and offspring tested for the loss of both genes. The double knockouts exhibit severe obesity as well as freezing behavior.

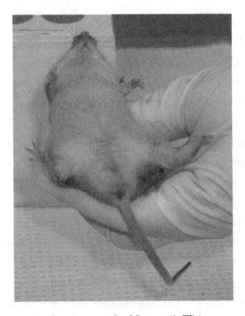

FIGURE 14: PC2-7B2 double knockout mouse in freezing mode. Note tail. This mouse is not anesthetized.

proteins in the biosynthesis of a wide variety of peptides [396]. Most recently, zebrafish PC2 has also been studied [397], and this organism could represent an interesting new model system to study the development of peptidergic signaling.

3.7 PC2 AS A THERAPEUTIC TARGET: DISEASE RELEVANCE

3.7.1 Polymorphisms

While clinical associations for the PC2 gene have lagged behind those of PC1/3, a small number of studies have now emerged. A variant of PC2 appears to be associated with susceptibility to type II diabetes in African Americans [398]; a different variant is associated with myocardial infarction in a Japanese population [399]. Specific 7B2 polymorphisms have also been associated with glucose levels in obese children [400]. It is interesting that for both PC1/3 and PC2, human polymorphisms are associated with aberrations in blood sugar control, indicating that pancreatic convertases play a large role in susceptibility to disease. However, more work needs to be performed to explore the role of PC2 in other diseases relevant to prohormone cleavage, for example in Cushing's disease [291] and thyroid disorders [189]. In the Goto-Kakizaki rat, a model of diabetes, PC2 expression was markedly reduced in islets, and the presumed PC2 substrate chromogranin A was also reduced [124]; these changes were found to be due to a reduction in the rate of biosynthesis of the enzyme under low glucose conditions, but did not impact proinsulin processing.

PC2 polymorphisms might manifest as amyloid disorders; knockdown of pancreatic PC2 led to impaired processing of human proislet amyloid polypeptide, which resulted in amyloid formation and cellular toxicity [401]. Interestingly, genetic deficiency for PC2 in mice is associated with protection from dietary fat-induced weight gain [397, 402]. This may confirm a quantum trait locus mapping study which indicated that PC2 is a candidate gene involved in body weight [403]. However, association between PC2 polymorphisms and obesity has not been observed in humans. This seems at variance with the clear association of the 7B2 gene with body weight homeostasis [368, 369] and supports the notion of physiological roles for 7B2 unrelated to PC2.

3.7.2 Cancer

Steiner and colleagues showed that prohormone convertase expression is often found in tumors of neuroendocrine origin [94]. PC2 is clearly expressed in pituitary adenomas, where it is associated with increased circulating levels of α-MSH [101, 404, 405 406]. However, expression of PC2 is higher in bronchial carcinoid tumors expressing ACTH than in pituitary adenomas expressing ACTH, indicating differential regulation in the two types of hypersteroidemia-causing tumors [407]. Expression of PC2 in lung carcinomas was also noted by Kajiwara *et al.* [408], Mbikay *et al.* [115], and North *et al.* [409]. PC2 has also been shown to be expressed in two breast cell cancer lines [410], but as yet has not been shown to be expressed in actual breast tumors [411]. Not surprisingly, human adrenomedullary tumors and phenochromocytomas also express PC2 [412, 413,

414], as do pancreatic tumors [415]. Interestingly, colorectal liver metastases express PC2 [266]. PC2 expression was also downregulated in thyroid carcinoma as compared to adenoma [416]. PC2 expression was reported in a human thymic tumor [417].

Many neuroendocrine tumors also express 7B2 (reviewed in [418]). An attempt has been made to use the presence of 7B2 antibodies as a diagnostic for nonfunctioning pituitary macroadenoma; a positive reaction for either anti-PC1/3 or anti-7B2 antibodies was significantly more frequent among patients with nonfunctioning pituitary macroadenoma than in other pituitary diseases and in healthy controls [265]. 7B2 was first suggested as a marker for neuroendocrine cancer as early as 1986 [294]; however, no increased levels of circulating 7B2 were detected in a recent study of patients with neuroendocrine tumors [419]. A high penetrance gene associated with colorectal cancer in the Ashkenazi population has been mapped to a region of chromosome 15 containing the 7B2 locus, and SNPs in this region are associated with increased risk of colorectal cancer [420]; however, this colorectal cancer risk locus has not yet been definitively mapped to the 7B2 gene itself.

CHAPTER 4

Summary and Future Directions

Progress on the prohormone convertases has continued, albeit more slowly, in the second decade following their discovery. One area which is currently experiencing significant research interest is the major mode of regulation of the convertases, particularly in disease states. The recent findings that oligomerized PC1/3 exhibits reduced enzymatic activity [28] and that the carboxy-tail domain of PC1/3 interacts with the catalytic domain [178] and with membranes [63] are of great interest in this regard, as is the demonstration that PC2 activity can be controlled by 7B2 levels [367]—which are themselves highly regulated by 5' regulatory elements. However, more work is required to associate alterations in PC regulation with various disease states. The increasing number of genome-wide association studies as well as other studies of obese patient populations promises to identify many new convertase/binding protein mutations and polymorphisms which affect catalytic activity and predisposition to disease. The biochemical studies which accompany the identification of such mutations will continue to shed light on the contribution of the various domains and specific residues to function. It appears likely that the involvement of PC variants in other diseases will become increasingly apparent as individual genomic sequencing becomes more common. Given recent data showing that the PC2 binding protein 7B2 is associated with blood sugar control [400], it will be especially interesting to learn whether variants of the PC1/3 binding protein proSAAS are associated with any clinical phenomena.

Another area which awaits further development is the production of potent and specific prohormone convertase-specific inhibitors. While initial attempts have been made to identify such inhibitors (M. Vivoli et al., in press [45, 172]), the medicinal chemistry efforts required to optimize weak leads are just now beginning. The discovery of PC-specific inhibitors would advance the study of the role of these enzymes in the various cleavage events involved in peptide hormone synthesis and could also be used to therapeutic advantage. For example, PC2 inhibitors would be expected to slow glucagon production while having relatively little effect on insulin production, a process potentially beneficial in diabetes. Recent technological advances in high-throughput screenings will no doubt yield additional inhibitor hits; improvements in selectivity, stability, and cell permeability will be important factors to consider during optimization steps. PC1/3 inhibitors could also be advantageous for crystallographic efforts by blocking the autocatalytic conversion to the smaller PC1/3 forms.

Clearly, having a crystal structure for the prohormone convertases would greatly speed our understanding of the regulation of these enzymes, the development of inhibitors, and clarify their interaction with both substrates and modulators. One of the most difficult challenges regarding their crystallization has been the propensity of these proteins to aggregate [28], an undesirable side reaction to the high concentrations required for crystallography. It should be noted that the only secretory granule protein which has been crystallized to date is peptidyl α-amidating enzyme [421]; like PC1/3, CPE also forms aggregates and has not been successfully crystallized after multiple attempts (L. Fricker, personal communication). It is interesting to note that prohormones—much more abundant residents of the granule compartment than convertases—have also proven refractory to crystallization, most likely due to the presence of flexible regions, such as the C-peptide in pro-insulin. The crystallization of sticky granule proteins such as the prohormone convertases remains a difficult future challenge. A ray of hope in this regard is that membrane proteins, once thought to be impossible to crystallize, are now being crystallized with regularity. Obtaining the structure of the proPC2–7B2 complex would be especially interesting, as it may help us to understand why proPC2 absolutely requires 7B2 to block spontaneous aggregation.

Lastly, increasing use should be made of knockout and mutant mice, as these models represent a valuable resource for understanding the contributions of PC1/3 and PC2 to the development of obesity and potentially to other diseases, for example, bone disease in the case of PC2. Further generation of PC mutant mice which recapitulate known human mutations, combined with studies which address the physiological effects of mutated convertase function, would provide valuable additional information on the contribution of convertase deficiency to human disease.

.   .   .   .

References

[1] Davidson HW, Peshavaria M, Hutton JC (1987) Proteolytic conversion of proinsulin into insulin. Identification of a Ca2+-dependent acidic endopeptidase in isolated insulin-secretory granules. *Biochem J* 246: 279–286.

[2] Davidson H, Rhodes C, Hutton J (1988) Intraorganellar calcium and pH control of proinsulin cleavage in the pancreatic beta cell via two distinct site-specific endopeptidases. *Nature* 333: 93–96.

[3] Lindberg I, Lincoln B, Rhodes CJ (1992) Fluorometric assay of a calcium-dependent, paired-basic processing endopeptidase present in insulinoma granules. *Biochem Biophys Res Commun* 183: 1–7.

[4] Smeekens SP, Steiner DF (1990) Identification of a human insulinoma cDNA encoding a novel mammalian protein structurally related to the yeast dibasic processing protease Kex2. *J Biol Chem* 265: 2997–3000.

[5] Ohagi S, LaMendola J, LeBeau MM, Espinosa R, 3rd, Takeda J, Smeekens SP, Chan SJ, Steiner DF (1992) Identification and analysis of the gene encoding human PC2, a prohormone convertase expressed in neuroendocrine tissues. *Proc Natl Acad Sci U S A* 89: 4977–4981.

[6] Smeekens SP, Avruch AS, LaMendola J, Chan SJ, Steiner DF (1991) Identification of a cDNA encoding a second putative prohormone convertase related to PC2 in AtT20 cells and islets of Langerhans. *Proc Natl Acad Sci U S A* 88: 340–344.

[7] Seidah NG, Marcinkiewicz M, Benjannet S, Gaspar L, Beaubien G, Mattei MG, Lazure C, Mbikay M, Chretien M (1991) Cloning and primary sequence of a mouse candidate prohormone convertase PC1 homologous to PC2, furin, and Kex2: distinct chromosomal localization and messenger RNA distribution in brain and pituitary compared to PC2. *Mol Endocrinol* 5: 111–122.

[8] Seidah NG, Hamelin J, Gaspar AM, Day R, Chretien M (1992) The cDNA sequence of the human pro-hormone and pro-protein convertase PC1. *DNA Cell Biol* 11: 283–289.

[9] Bailyes E, Shennan KIJ, Seal AJ, Smeekens SP, Steiner DF, Hutton JC, Docherty K (1992) A member of the eukaryotic subtilisin family(PC3) has the enzymic properties of the type 1 proinsulin-converting endopeptidase. *Biochem J* 285: 391–394.

[10] Bennett DL, Bailyes EM, Nielsen E, Guest PC, Rutherford NG, Arden SD, Hutton JC (1992) Identification of the type 2 proinsulin processing endopeptidase as PC2, a member of the eukaryotic subtilisin family. *J Biol Chem* 267: 15229–15236.

[11] Marcinkiewicz M, Seidah NG, Chretien M (1996) Implications of the subtilisin/kexin-like precursor convertases in the development and function of nervous tissues. *Acta Neurobiol Exp (Wars)* 56: 287–298.

[12] Seidah NG, Chretien M, Day R (1994) The family of subtilisin/kexin like pro-protein and pro-hormone convertases: divergent or shared functions. *Biochimie* 76: 197–209.

[13] Torii S, Yamagishi T, Murakami K, Nakayama K (1993) Localization of Kex2-like processing endoproteases, furin and PC4, within mouse testis by in situ hybridization. *FEBS Lett* 316: 12–16.

[14] Kiefer MC, Tucker JE, Joh R, Landsberg KE, Saltman D, Barr PJ (1991) Identification of a second human subtilisin-like protease gene in the fes/fps region of chromosome 15. *DNA Cell Biol* 10: 757–769.

[15] Seidah NG, Hamelin J, Mamarbachi M, Dong W, Tardos H, Mbikay M, Chretien M, Day R (1996) cDNA structure, tissue distribution, and chromosomal localization of rat PC7, a novel mammalian proprotein convertase closest to yeast kexin-like proteinases. *Proc Natl Acad Sci U S A* 93: 3388–3393.

[16] Nakayama K, Kim WS, Torii S, Hosaka M, Nakagawa T, Ikemizu J, Baba T, Murakami K (1992) Identification of the fourth member of the mammalian endoprotease family homologous to the yeast Kex2 protease. *J Biol Chem* 267: 5897–5900.

[17] Seidah NG, Day R, Hamelin J, Gaspar A, Collard MW, Chretien M (1992) Testicular expression of PC4 in the rat: molecular diversity of a novel germ cell-specific Kex2/subtilisin-like proprotein convertase. *Mol Endocrinol* 6: 1559–1570.

[18] Lusson J, Vieau D, Hamelin J, Day R, Chretien M, Seidah NG (1993) cDNA structure of the mouse and rat subtilisin/kexin-like PC5: a candidate proprotein convertase expressed in endocrine and nonendocrine cells. *Proc Natl Acad Sci U S A* 90: 6691–6695.

[19] Nakagawa T, Murakami K, Nakayama K (1993) Identification of an isoform with an extremely large Cys-rich region of PC6, a Kex2-like processing endoprotease. *FEBS Lett* 327: 165–171.

[20] Seidah NG, Mowla SJ, Hamelin J, Mamarbachi AM, Benjannet S, Toure BB, Basak A, Munzer JS, Marcinkiewicz J, Zhong M, Barale JC, Lazure C, Murphy RA, Chretien M, Marcinkiewicz M (1999) Mammalian subtilisin/kexin isozyme SKI-1: A widely expressed proprotein convertase with a unique cleavage specificity and cellular localization. *Proc Natl Acad Sci U S A* 96: 1321–1326.

[21] Seidah NG, Benjannet S, Wickham L, Marcinkiewicz J, Jasmin SB, Stifani S, Basak A, Prat A, Chretien M (2003) The secretory proprotein convertase neural apoptosis-regulated

convertase 1 (NARC-1): liver regeneration and neuronal differentiation. *Proc Natl Acad Sci U S A* 100: 928–933.

[22]   Dupuy A, Lindberg I, Zhou Y, Akil H, Lazure C, Chretien M, Seidah NG, Day R (1994) Processing of prodynorphin by the prohormone convertase PC1 results in high molecular weight intermediate forms. Cleavage at a single arginine residue. *FEBS Lett* 337: 60–65.

[23]   Friedman TC, Loh YP, Birch NP (1994) In vitro processing of proopiomelanocortin by recombinant PC1 (SPC3). *Endocrinology* 135: 854–862.

[24]   Zhou Y, Lindberg I (1993) Purification and characterization of the prohormone convertase PC1(PC3). *J Biol Chem* 268: 5615–5623.

[25]   Rufaut NW, Brennan SO, Hakes DJ, Dixon JE, Birch NP (1993) Purification and characterization of the candidate prohormone-processing enzyme SPC3 produced in a mouse L cell line. *J Biol Chem* 268: 20291–20298.

[26]   Boudreault A, Gauthier D, Rondeau N, Savaria D, Seidah NG, Chretien M, Lazure C (1998) Molecular characterization, enzymatic analysis, and purification of murine proprotein convertase-1/3 (PC1/PC3) secreted from recombinant baculovirus-infected insect cells. *Protein Expr Purif* 14: 353–366.

[27]   Zhou Y, Lindberg I (1994) Enzymatic properties of carboxyl-terminally truncated prohormone convertase 1 (PC1/SPC3) and evidence for autocatalytic conversion. *J Biol Chem* 269: 18408–18413.

[28]   Hoshino A, Kowalska D, Jean F, Lazure C, Lindberg I (2011) Modulation of PC1/3 Activity by Self-Interaction and Substrate Binding. *Endocrinology* 152: 1402–1411.

[29]   Oliva AA, Jr., Steiner DF, Chan SJ (1995) Proprotein convertases in amphioxus: predicted structure and expression of proteases SPC2 and SPC3. *Proc Natl Acad Sci U S A* 92: 3591–3595.

[30]   Ruoslahti E, Pierschbacher MD (1986) Arg-Gly-Asp: a versatile cell recognition signal. *Cell* 44: 517–518.

[31]   Lusson J, Benjannet S, Hamelin J, Savaria D, Chretien M (1997) The integrity of the RRGDL sequence of the proprotein convertase PC1 is critical for its zymogen and C-terminal processing and for its cellular trafficking. *Biochem J* 326: 737–744.

[32]   Munzer JS, Basak A, Zhong M, Mamarbachi A, Hamelin J, Savaria D, Lazure C, Benjannet S, Chretien M, Seidah NG (1997) In vitro characterization of the novel proprotein convertase PC7. *J Biol Chem* 272: 19672–19681.

[33]   Chun JY, Korner J, Kreiner T, Scheller RH, Axel R (1994) The function and differential sorting of a family of aplysia prohormone processing enzymes. *Neuron* 12: 831–844.

[34]   Morash MG, MacDonald AB, Croll RP, Anini Y (2009) Molecular cloning, ontogeny and tissue distribution of zebrafish (Danio rerio) prohormone convertases: pcsk1 and pcsk2. *Gen Comp Endocrinol* 162: 179–187.

[35]   Gorham EL, Nagle GT, Smith JS, Shen H, Kurosky A (1996) Molecular cloning of pro-hormone convertase 1 from the atrial gland of Aplysia. *DNA Cell Biol* 15: 339–345.

[36]   Chan SJ, Oliva AAJ, LaMendola J, Grens A, Bode H, Steiner DF (1992) Conservation of the prohormone convertase gene family in metazoa: analysis of cDNAs encoding a PC3-like protein from hydra. *Proc Natl Acad Sci U S A* 89: 6678–6682.

[37]   Seidah NG, Chretien M (1994) Pro-protein convertases of the subtilisin/kexin family. *Methods Enzymol* 244: 171–188.

[38]   Siezen R, Leunissen JA (1997) Subtilases: the superfamily of subtilisin-like serine prote-ases. *Protein Sci* 6: 501–523.

[39]   Tangrea MA, Alexander P, Bryan PN, Eisenstein E, Toedt J, Orban J (2001) Stability and global fold of the mouse prohormone convertase 1 pro-domain. *Biochemistry* 40: 5488–5495.

[40]   Baker D, Shiau AK, Agard DA (1993) The role of pro regions in protein folding. *Curr Opin Cell Biol* 5: 966–970.

[41]   Shinde U, Thomas G (2011) Insights from bacterial subtilases into the mechanisms of intramolecular chaperone-mediated activation of furin. *Methods Mol Biol* 768: 59–06.

[42]   Muller L, Cameron A, Fortenberry Y, Apletalina EV, Lindberg I (2000) Processing and sorting of the prohormone convertase 2 propeptide. *J Biol Chem* 275: 39213–39222.

[43]   Basak A, Lazure C (2003) Synthetic peptides derived from the prosegments of proprotein convertase 1/3 and furin are potent inhibitors of both enzymes. *Biochem J* 373: 231–239.

[44]   Rabah N, Gauthier D, Wilkes BC, Gauthier DJ, Lazure C (2006) Single amino acid substi-tution in the PC1/3 propeptide can induce significant modifications of its inhibitory profile toward its cognate enzyme. *J Biol Chem* 281: 7556–7567.

[45]   Fugere M, Limperis PC, Beaulieu-Audy V, Gagnon F, Lavigne P, Klarskov K, Leduc R, Day R (2002) Inhibitory potency and specificity of subtilase-like pro-protein convertase (SPC) prodomains. *J Biol Chem* 277: 7648–7656.

[46]   Inouye M (1991) Intramolecular chaperones: the role of the pro-peptide in protein folding. *Enzyme* 45: 314–321.

[47]   Powner D, J D (1998) Activation of the kexin from *Schizosaccharomyces pombe* requires in-ternal cleavage of its initial cleaved prosequence. *Mol Cell Biol* 18: 400–408.

[48]   Lesage G, Tremblay M, Guimond J, Boileau G (2001) Mechanism of Kex2p inhibition by its proregion. *FEBS Lett* 508: 332–336.

[49]   Goodman LJ, Gorman CM (1994) Autoproteolytic activation of the mouse prohormone convertase mPC1. *Biochem Biophys Res Commun* 201: 795–804.

[50]   Zhou A, Paquet L, Mains RE (1995) Structural elements that direct specific process-ing of different mammalian subtilisin-like prohormone convertases. *J Biol Chem* 270: 21509–21516.

[51] Tangrea MA, Bryan PN, Sari N, Orban J (2002) Solution structure of the pro-hormone convertase 1 pro-domain from Mus musculus. *J Mol Biol* 320: 801–812.

[52] Lipkind G, Gong Q, Steiner DF (1995) Molecular modeling of the substrate specificity of prohormone convertases SPC2 and SPC3. *J Biol Chem* 270: 13277–13284.

[53] Steiner DF, Smeekens SP, Ohagi S, Chan SJ (1992) The new enzymology of precursor processing endoproteases. *J Biol Chem* 267: 23435–23438.

[54] Henrich S, Cameron A, Bourenkov GP, Kiefersauer R, Huber R, Lindberg I, Bode W, Than ME (2003) The crystal structure of the proprotein processing proteinase furin explains its stringent specificity. *Nat Struct Biol* 10: 520–526.

[55] Jackson RS, Creemers JW, Ohagi S, Raffin-Sanson ML, Sanders L, Montague CT, Hutton JC, O'Rahilly S (1997) Obesity and impaired prohormone processing associated with mutations in the human prohormone convertase 1 gene. *Nat Genet* 16: 303–306.

[56] Rovere C, Luis J, Lissitzky JC, Basak A, Marvaldi J, Chretien M, Seidah NG (1999) The RGD motif and the C-terminal segment of proprotein convertase 1 are critical for its cellular trafficking but not for its intracellular binding to integrin alpha5beta1. *J Biol Chem* 274: 12461–12467.

[57] Ueda K, Lipkind GM, Zhou A, Zhu X, Kuznetsov A, Philipson L, Gardner P, Zhang C, Steiner DF (2003) Mutational analysis of predicted interactions between the catalytic and P domains of prohormone convertase 3 (PC3/PC1). *Proc Natl Acad Sci U S A* 100: 5622–5627.

[58] Zhou A, Martin S, Lipkind G, LaMendola J, Steiner DF (1998) Regulatory role of the P domain of subtilisin-like prohormone convertases. *J Biol Chem* 273: 11107–11114.

[59] Lipkind GM, Zhou A, Steiner DF (1998) A model for the structure of the P domain in the subtilisin-like prohormone convertases. *Proc Natl Acad Sci U S A* 95: 7310–7315.

[60] Jutras I, Seidah NG, Reudelhuber TL, Brechler V (1997) Two activation states of the prohormone convertase PC1 in the secretory pathway. *J Biol Chem* 272: 15184–15188.

[61] Jutras I, Seidah NG, Reudelhuber TL (2000) A predicted alpha -helix mediates targeting of the proprotein convertase PC1 to the regulated secretory pathway. *J Biol Chem* 275: 40337–40343.

[62] Bernard N, Kitabgi P, Rovere-Jovene C (2003) The Arg617-Arg618 cleavage site in the C-terminal domain of PC1 plays a major role in the processing and targeting of the enzyme within the regulated secretory pathway. *J Neurochem* 85: 1592–1603.

[63] Dikeakos JD, Di Lello P, Lacombe MJ, Ghirlando R, Legault P, Reudelhuber TL, Omichinski JG (2009) Functional and structural characterization of a dense core secretory granule sorting domain from the PC1/3 protease. *Proc Natl Acad Sci U S A* 106: 7408–7413.

[64] Dikeakos JD, Mercure C, Lacombe MJ, Seidah NG, Reudelhuber TL (2007) PC1/3,

PC2 and PC5/6A are targeted to dense core secretory granules by a common mechanism. *FEBS J* 274: 4094–4102.

[65] Lou H, Smith AM, Coates LC, Cawley NX, Loh YP, Birch NP (2007) The transmembrane domain of the prohormone convertase PC3: a key motif for targeting to the regulated secretory pathway. *Mol Cell Endocrinol* 267: 17–25.

[66] Benjannet S, Reudelhuber T, Mercure C, Rondeau N, Chretien M, Seidah NG (1992) Proprotein conversion is determined by a multiplicity of factors including convertase processing, substrate specificity, and intracellular environment. Cell type-specific processing of human prorenin by the convertase PC1. *J Biol Chem* 267: 11417–11423.

[67] Zhou A, Mains RE (1994) Endoproteolytic processing of proopiomelanocortin and prohormone convertases 1 and 2 in neuroendocrine cells overexpressing prohormone convertases 1 or 2. *J Biol Chem* 269: 17440–17447.

[68] Lindberg I, Ahn SC, Breslin MB (1994) Cellular distributions of the prohormone processing enzymes PC1 and PC2. *Mol Cell Neurosci* 5: 614–622.

[69] Milgram SL, Mains RE (1994) Differential effects of temperature blockade on the proteolytic processing of three secretory granule-associated proteins. *J Cell Sci.* pp. 737–745.

[70] Vindrola O, Lindberg I (1992) Biosynthesis of the prohormone convertase mPC1 in AtT-20 cells. *Mol Endocrinol* 6: 1088–1094.

[71] Fricker LD, McKinzie AA, Sun J, Curran E, Qian Y, Yan L, Patterson SD, Courchesne PL, Richards B, Levin N, Mzhavia N, Devi LA, Douglass J (2000) Identification and characterization of proSAAS, a granin-like neuroendocrine peptide precursor that inhibits prohormone processing. *J Neurosci* 20: 639–648.

[72] Bures EJ, Courchesne PL, Douglass J, Chen K, Davis MT, Jones MD, McGinley MD, Robinson JH, Spahr CS, Sun J, Wahl RC, Patterson SD (2001) Identification of incompletely processed potential carboxypeptidase E substrates from CpEfat/CpEfat mice. *Proteomics* 1: 79–92.

[73] Bartolomucci A, Possenti R, Mahata SK, Fischer-Colbrie R, Loh YP, Salton SR (2011) The Extended Granin Family: Structure, Function, and Biomedical Implications. *Endocr Rev* 32: 755–797.

[74] Kudo H, Liu J, Jansen EJ, Ozawa A, Panula P, Martens GJ, Lindberg I (2009) Identification of proSAAS homologs in lower vertebrates: conservation of hydrophobic helices and convertase-inhibiting sequences. *Endocrinology* 150: 1393–1399.

[75] Qian Y, Devi LA, Mzhavia N, Munzer S, Seidah NG, Fricker LD (2000) The C-terminal region of proSAAS is a potent inhibitor of prohormone convertase 1. *J Biol Chem* 275: 23596–23601.

[76] Cameron A, Fortenberry Y, Lindberg I (2000) The SAAS granin exhibits structural and

functional homology to 7B2 and contains a highly potent hexapeptide inhibitor of PC1. *FEBS Lett* 473: 135–138.

[77] Morgan DJ, Mzhavia N, Peng B, Pan H, Devi LA, Pintar JE (2005) Embryonic gene expression and pro-protein processing of proSAAS during rodent development. *J Neurochem* 93: 1454–1462.

[78] Apletalina E, Appel J, Lamango NS, Houghten RA, Lindberg I (1998) Identification of inhibitors of prohormone convertases 1 and 2 using a peptide combinatorial library. *J Biol Chem* 273: 26589–26595.

[79] Atkins N, Jr., Mitchell JW, Romanova EV, Morgan DJ, Cominski TP, Ecker JL, Pintar JE, Sweedler JV, Gillette MU (2010) Circadian integration of glutamatergic signals by little SAAS in novel suprachiasmatic circuits. *PLoS One* 5: e12612.

[80] Morgan DJ, Wei S, Gomes I, Czyzyk T, Mzhavia N, Pan H, Devi LA, Fricker LD, Pintar JE (2010) The propeptide precursor proSAAS is involved in fetal neuropeptide processing and body weight regulation. *J Neurochem* 113: 1275–1284.

[81] Wardman JH, Berezniuk I, Di S, G. TJ, Fricker LD (2011) ProSAAS-derived peptides are colocalized with Neuropeptide Y and function as neuropeptides in the regulation of food intake. *PLoS One* 6.

[82] Kikuchi K, Arawaka S, Koyama S, Kimura H, Ren CH, Wada M, Kawanami T, Kurita K, Daimon M, Kawakatsu S, Kadoya T, Goto K, Kato T (2003) An N-terminal fragment of ProSAAS (a granin-like neuroendocrine peptide precursor) is associated with tau inclusions in Pick's disease. *Biochem Biophys Res Commun* 308: 646–654.

[83] Wada M, Ren CH, Koyama S, Arawaka S, Kawakatsu S, Kimura H, Nagasawa H, Kawanami T, Kurita K, Daimon M, Hirano A, Kato T (2004) A human granin-like neuroendocrine peptide precursor (proSAAS) immunoreactivity in tau inclusions of Alzheimer's disease and parkinsonism-dementia complex on Guam. *Neurosci Lett* 356: 49–52.

[84] Jahn H, Wittke S, Zurbig P, Raedler TJ, Arlt S, Kellmann M, Mullen W, Eichenlaub M, Mischak H, Wiedemann K (2011) Peptide fingerprinting of Alzheimer's disease in cerebrospinal fluid: identification and prospective evaluation of new synaptic biomarkers. *PLoS One* 6: e26540.

[85] Finehout EJ, Franck Z, Choe LH, Relkin N, Lee KH (2007) Cerebrospinal fluid proteomic biomarkers for Alzheimer's disease. *Ann Neurol* 61: 120–129.

[86] Pasinetti GM, Ungar LH, Lange DJ, Yemul S, Deng H, Yuan X, Brown RH, Cudkowicz ME, Newhall K, Peskind E, Marcus S, Ho L (2006) Identification of potential CSF biomarkers in ALS. *Neurology* 66: 1218–1222.

[87] Seidah NG, Mattei MG, Gaspar L, Benjannet S, Mbikay M, Chretien M (1991) Chromosomal assignments of the genes for neuroendocrine convertase PC1 (NEC1) to human

5q15–21, neuroendocrine convertase PC2 (NEC2) to human 20p11.1-11.2, and furin (mouse 7[D1-E2] region). *Genomics* 11: 103–107.

[88] Ftouhi N, Day R, Mbikay M, Chretien M, Seidah NG (1994) Gene organization of the mouse pro-hormone and pro-protein convertase PC1. *DNA Cell Biol* 13: 395–407.

[89] Szpirer C, Tissir F, Riviere M, Van Vooren P, Kela J, Lallemand F, Gabant P, Hoebee B, Klinga-Levan K, Levan G, Szpirer J (1999) Rat Chromosome 2: assignment of the genes encoding cyclin B1, interleukin 6 signal transducer, and proprotein convertase 1 to the Mcs1-containing region and identification of new microsatellite markers. *Mamm Genome* 10: 30–34.

[90] Quan X, Laes JF, Ravoet M, Van Vooren P, Szpirer J, Szpirer C (2000) Localization of new, microdissection- generated, anonymous markers and of the genes Pcsk1, Dhfr, Ndub13, and Ccnb1 to rat chromosome region 2q1. *Cytogenet Cell Genet* 88: 119–123.

[91] Cullinan WE, Day NC, Schafer MK, Day R, Seidah NG, Chretien M, Akil H, Watson SJ (1991) Neuroanatomical and functional studies of peptide precursor-processing enzymes. *Enzyme* 45: 285–300.

[92] Lanoue E, Day R (2001) Coexpression of proprotein convertase SPC3 and the neuroendo-crine precursor proSAAS. *Endocrinology* 142: 4141–4149.

[93] Day R, Schafer MK, Watson SJ, Chretien M, Seidah NG (1992) Distribution and regula-tion of the prohormone convertases PC1 and PC2 in the rat pituitary. *Mol Endocrinol* 6: 485–497.

[94] Scopsi L, Gullo M, Rilke F, Martin S, Steiner DF (1995) Proprotein convertases (PC1/PC3 and PC2) in normal and neoplastic human tissues: their use as markers of neuroendo-crine differentiation. *J Clin Endocrinol Metab* 80: 294–301.

[95] Seidah NG, Day R, Benjannet S, Rondeau N, Boudreault A, Reudelhuber T, Schafer MK, Watson SJ, Chretien M (1992) The prohormone and proprotein processing enzymes PC1 and PC2: structure, selective cleavage of mouse POMC and human renin at pairs of basic residues, cellular expression, tissue distribution, and mRNA regulation. *NIDA Res Monogr* 126: 132–150.

[96] Schafer MK, Day R, Cullinan WE, Chretien M, Seidah NG, Watson SJ (1993) Gene ex-pression of prohormone and proprotein convertases in the rat CNS: a comparative in situ hybridization analysis. *J Neurosci* 13: 1258–1279.

[97] Dong W, Seidel B, Marcinkiewicz M, Chretien M, Seidah NG, Day R (1997) Cellular localization of the prohormone convertases in the hypothalamic paraventricular and supra-optic nuclei: selective regulation of PC1 in corticotrophin-releasing hormone parvocellular neurons mediated by glucocorticoids. *J Neurosci* 17: 563–575.

[98] Billova S, Galanopoulou AS, Seidah NG, Qiu X, Kumar U (2007) Immunohistochemical

expression and colocalization of somatostatin, carboxypeptidase-E and prohormone convertases 1 and 2 in rat brain. *Neuroscience* 147: 403–418.

[99] Feng Y, Reznik SE, Fricker LD (2001) Distribution of proSAAS-derived peptides in rat neuroendocrine tissues. *Neuroscience* 105: 469–478.

[100] Fuller JA, Brun-Zinkernagel AM, Clark AF, Wordinger RJ (2009) Subtilisin-like proprotein convertase expression, localization, and activity in the human retina and optic nerve head. *Invest Ophthalmol Vis Sci* 50: 5759–5768.

[101] Takumi I, Steiner DF, Sanno N, Teramoto A, Osamura RY (1998) Localization of prohormone convertases 1/3 and 2 in the human pituitary gland and pituitary adenomas: analysis by immunohistochemistry, immunoelectron microscopy, and laser scanning microscopy. *Mod Pathol* 11: 232–238.

[102] Tanaka S, Kurabuchi S, Mochida H, Kato T, Takahashi S, Watanabe T, Nakayama K (1996) Immunocytochemical localization of prohormone convertases PC1/PC3 and PC2 in rat pancreatic islets. *Arch Histol Cytol* 59: 261–271.

[103] Itoh Y, Tanaka S, Takekoshi S, Itoh J, Osamura RY (1996) Prohormone convertases (PC1/3 and PC2) in rat and human pancreas and islet cell tumors: subcellular immunohistochemical analysis. *Pathol Int* 46: 726–737.

[104] Damholt AB, Buchan AM, Holst JJ, Kofod H (1999) Proglucagon processing profile in canine L cells expressing endogenous prohormone convertase 1/3 and prohormone convertase 2. *Endocrinology* 140: 4800–4808.

[105] Kurabuchi S, Tanaka S (2002) Immunocytochemical localization of prohormone convertases PC1 and PC2 in the mouse thyroid gland and respiratory tract. *J Histochem Cytochem* 50: 903–909.

[106] Tomita T (2000) Immunocytochemical Localization of Prohormone Convertase 1/3 and 2 in Thyroid C-Cells and Medullary Thyroid Carcinomas. *Endocr Pathol* 11: 165–172.

[107] McGirr R, Ejbick CE, Carter DE, Andrews JD, Nie Y, Friedman TC, Dhanvantari S (2005) Glucose dependence of the regulated secretory pathway in alphaTC1-6 cells. *Endocrinology* 146: 4514–4523.

[108] Whalley NM, Pritchard LE, Smith DM, White A (2011) Processing of proglucagon to GLP-1 in pancreatic alpha-cells: is this a paracrine mechanism enabling GLP-1 to act on beta-cells? *J Endocrinol* 211: 99–106.

[109] Portela-Gomes GM, Grimelius L, Stridsberg M (2008) Prohormone convertases 1/3, 2, furin and protein 7B2 (Secretogranin V) in endocrine cells of the human pancreas. *Regul Pept* 146: 117–124.

[110] Slominski A, Wortsman J (2003) Self-regulated endocrine systems in the skin. *Minerva Endocrinol* 28: 135–143.

[111]   Sharp BM, McKean DJ, McAllen K, Shahabi NA (1998) Signaling through delta opioid receptors on murine splenic T cells and stably transfected Jurkat cells. *Ann N Y Acad Sci* 840: 420–424.

[112]   Philippe D, Dubuquoy L, Groux H, Brun V, Chuoi-Mariot MT, Gaveriaux-Ruff C, Colombel JF, Kieffer BL, Desreumaux P (2003) Anti-inflammatory properties of the mu opioid receptor support its use in the treatment of colon inflammation. *J Clin Invest* 111: 1329–1338.

[113]   Krajnik M, Schafer M, Sobanski P, Kowalewski J, Bloch-Boguslawska E, Zylicz Z, Mousa SA (2010) Enkephalin, its precursor, processing enzymes, and receptor as part of a local opioid network throughout the respiratory system of lung cancer patients. *Hum Pathol* 41: 632–642.

[114]   Creemers JW, Roebroek AJ, Van de Ven WJ (1992) Expression in human lung tumor cells of the proprotein processing enzyme PC1/PC3. Cloning and primary sequence of a 5 kb cDNA. *FEBS Lett* 300: 82–88.

[115]   Mbikay M, Sirois F, Yao J, Seidah NG, Chretien M (1997) Comparative analysis of expression of the proprotein convertases furin, PACE4, PC1 and PC2 in human lung tumours. *Br J Cancer* 75: 1509–1514.

[116]   Vindrola O, Mayer AM, Citera G, Spitzer JA, Espinoza LR (1994) Prohormone convertases PC2 and PC3 in rat neutrophils and macrophages. Parallel changes with proenkephalin-derived peptides induced by LPS in vivo. *Neuropeptides* 27: 235–244.

[117]   LaMendola J, Martin SK, Steiner DF (1997) Expression of PC3, carboxypeptidase E and enkephalin in human monocyte-derived macrophages as a tool for genetic studies. *FEBS Lett* 404: 19–22.

[118]   Horsch D, Day R, Seidah NG, Weihe E, Schafer MK (1997) Immunohistochemical localization of the pro-peptide processing enzymes PC1/PC3 and PC2 in the human anal canal. *Peptides* 18: 755–760.

[119]   Mzhavia N, Berman Y, Che FY, Fricker LD, Devi LA (2001) ProSAAS processing in mouse brain and pituitary. *J Biol Chem* 276: 6207–6213.

[120]   Sayah M, Fortenberry Y, Cameron A, Lindberg I (2001) Tissue distribution and processing of proSAAS by proprotein convertases. *J Neurochem* 76: 1833–1841.

[121]   Hatcher NG, Atkins NJ, Annangudi SP, Forbes AJ, Kelleher NL, Gillette MU, Sweedler JV (2008) Mass spectrometry-based discovery of circadian peptides. *Proc Natl Acad Sci U S A* 105: 12527–12532.

[122]   Lee JE, Atkins N, Jr., Hatcher NG, Zamdborg L, Gillette MU, Sweedler JV, Kelleher NL (2010) Endogenous peptide discovery of the rat circadian clock: a focused study of the suprachiasmatic nucleus by ultrahigh performance tandem mass spectrometry. *Mol Cell Proteomics* 9: 285–297.

[123] Feng S, Chen JK, Yu H, Simon JA, Schreiber SL (1994) Two binding orientations for peptides to the Src SH3 domain developement of a general model for SH3-ligand interactions. *Science* 266: 1241–1246.

[124] Guest PC, Abdel-Halim SM, Gross DJ, Clark A, Poitout V, Amaria R, Ostenson CG, Hutton JC (2002) Proinsulin processing in the diabetic Goto-Kakizaki rat. *J Endocrinol* 175: 637–647.

[125] Tanaka S, Yora T, Nakayama K, Inoue K, Kurosumi K (1997) Proteolytic processing of proopiomelanocortin occurs in acidfying secretory granules of AtT-20 cells. *J Histochem Cytochem* 45: 425–436.

[126] Arias AE, Velez-Granell CS, Mayer G, Bendayan M (2000) Colocalization of chaperone Cpn60, proinsulin and convertase PC1 within immature secretory granules of insulin-secreting cells suggests a role for Cpn60 in insulin processing. *J Cell Sci* 113 (Pt 11): 2075–2083.

[127] Hornby PJ, Rosenthal SD, Mathis JP, Vindrola O, Lindberg I (1993) Immunocytochemical localization of the neuropeptide-synthesizing enzyme PC1 in AtT-20 cells. *Neuroendocrinology* 58: 555–563.

[128] Paquet L, Zhou A, Chang EY, Mains RE (1996) Peptide biosynthetic processing: distinguishing prohormone convertases PC1 and PC2. *Mol Cell Endocrinol* 120: 161–168.

[129] Bloomquist BT, Eipper BA, Mains RE (1991) Prohormone-converting enzymes: regulation and evaluation of function using antisense RNA. *Mol Endocrinol* 5: 2014–2024.

[130] Varlamov O, Fricker LD, Furukawa H, Steiner DF, Langley SH, Leiter EH (1997) Beta-cell lines derived from transgenic Cpe(fat)/Cpe(fat) mice are defective in carboxypeptidase E and proinsulin processing. *Endocrinology* 138: 4883–4892.

[131] Ugleholdt R, Poulsen ML, Holst PJ, Irminger JC, Orskov C, Pedersen J, Rosenkilde MM, Zhu X, Steiner DF, Holst JJ (2006) Prohormone convertase 1/3 is essential for processing of the glucose-dependent insulinotropic polypeptide precursor. *J Biol Chem* 281: 11050–11057.

[132] Cain BM, Vishnuvardhan D, Beinfeld MC (2001) Neuronal cell lines expressing PC5, but not PC1 or PC2, process Pro-CCK into glycine-extended CCK 12 and 22. *Peptides* 22: 1271–1277.

[133] Marandi M, Mowla SJ, Tavallaei M, Yaghoobi MM, Jafarnejad SM (2007) Proprotein convertases 1 and 2 (PC1 and PC2) are expressed in neurally differentiated rat bone marrow stromal stem cells (BMSCs). *Neurosci Lett* 420: 198–203.

[134] Beinfeld MC, Wang W (2002) CCK processing by pituitary GH3 cells, human teratocarcinoma cells NT2 and hNT differentiated human neuronal cells evidence for a differentiation-induced change in enzyme expression and pro CCK processing. *Life Sci* 70: 1251–1258.

[135] St Germain C, Croissandeau G, Mayne J, Baltz JM, Chretien M, Mbikay M (2005) Expression and transient nuclear translocation of proprotein convertase 1 (PC1) during mouse preimplantation embryonic development. *Mol Reprod Dev* 72: 483–493.

[136] Feng Y, Reznik SE, Fricker LD (2002) ProSAAS and prohormone convertase 1 are broadly expressed during mouse development. *Brain Res Gene Expr Patterns* 1: 135–140.

[137] Marcinkiewicz M, Day R, Seidah NG, Chretien M (1993) Ontogeny of the prohormone convertases PC1 and PC2 in the mouse hypophysis and their colocalization with corticotropin and alpha- melanotropin. *Proc Natl Acad Sci U S A* 90: 4922–4926.

[138] Zheng M, Streck RD, Scott RE, Seidah NG, Pintar JE (1994) The developmental expression in rat of proteases furin, PC1, PC2, and carboxypeptidase E: implications for early maturation of proteolytic processing capacity. *J Neurosci* 14: 4656–4673.

[139] Lindberg I (1994) Evidence for cleavage of the PC1/PC3 pro-segment in the endoplasmic reticulum. *Mol Cell Neurosci* 5: 263–268.

[140] Anderson ED, VanSlyke JK, Thulin CD, Jean F, Thomas G (1997) Activation of the furin endoprotease is a multiple-step process: requirements for acidification and internal propeptide cleavage. *EMBO J* 16: 1508–1518.

[141] Gluschankof P, Fuller RS (1994) A C-terminal domain conserved in precursor processing proteases is required for intramolecular N-terminal maturation of proKex2 protease. *EMBO J* 13: 2280–2288.

[142] Creemers JWM, Siezen RJ, Roebroek AJM, Ayoubi TAY, Huylebroek D, WJM vdV (1993) Modulation of furin-mediated proprotein processing activity by site-directed mutagenesis. *J Biol Chem* 268: 21826–21834.

[143] Scougall K, Taylor NA, Jermany JL, Docherty K, Shennan KI (1998) Differences in the autocatalytic cleavage of pro-PC2 and pro-PC3 can be attributed to sequences within the propeptide and Asp310 of pro-PC2. Biochem J 334 (Pt 3): 531–537.

[144] Lee SN, Prodhomme E, Lindberg I (2004) Prohormone convertase 1 (PC1) processing and sorting: effect of PC1 propeptide and proSAAS. *J Endocrinol* 182: 353–364.

[145] Benjannet S, Rondeau N, Paquet L, Boudreault A, Lazure C, Chretien M, Seidah NG (1993) Comparative biosynthesis, covalent post-translational modifications and efficiency of prosegment cleavage of the prohormone convertases PC1 and PC2: glycosylation, sulphation and identification of the intracellular site of prosegment cleavage of PC1 and PC2. *Biochem J* 294 (Pt 3): 735–743.

[146] Shennan KIJ, Taylor NA, Jermany JL, Matthews G, Docherty K (1995) Differences in pH optima and calcium requirements for maturation of the prohormone convertases PC2 and PC3 indicates different intracellular locations for these events. *J Biol Chem* 270: 1402–1407.

[147] Zandberg WF, Benjannet S, Hamelin J, Pinto BM, Seidah NG (2011) N-Glycosylation controls trafficking, zymogen activation, and substrate processing of proprotein convertases PC1/3 and SKI-1. *Glycobiology* 21: 1290–1300.

[148] Zhou A, Bloomquist BT, Mains RE (1993) The prohormone convertases PC1 and PC2 mediate distinct endoproteolytic cleavages in a strict temporal order during proopiomelanocortin biosynthetic processing. *J Biol Chem* 268: 1763–1769.

[149] Fernandez CJ, Haugwitz M, Eaton B, Moore HP (1997) Distinct molecular events during secretory granule biogenesis revealed by sensitivities to brefeldin A. *Mol Biol Cell* 8: 2171–2185.

[150] Zhou Y, Rovere C, Kitabgi P, Lindberg I (1995) Mutational analysis of PC1 (SPC3) in PC12 cells. 66-kDa PC1 is fully functional. *J Biol Chem* 270: 24702–24706.

[151] Coates LC, Birch NP (1997) Posttranslational maturation of the prohormone convertase sPC3 in vitro. *J Neurochem* 68: 828–836.

[152] Arvan P, Castle D (1998) Sorting and storage during secretory granule biogenesis: looking backward and looking forward. *Biochem J* 332 (Pt 3): 593–610.

[153] Dikeakos JD, Lacombe MJ, Mercure C, Mireuta M, Reudelhuber TL (2007) A hydrophobic patch in a charged alpha-helix is sufficient to target proteins to dense core secretory granules. *J Biol Chem* 282: 1136–1143.

[154] Hill RM, Ledgerwood EC, Brennan SO, Loh YP, Christie DL, Bich NP (1995) Comparison of the molecular forms of the kex2/subtilisin-like serine proteases SPC2, SPC3 and furin in neuroendocrine secretory vesicles reveals differences in carboxyl-terminus truncation and membrane association. *J Neurochem* 65: 2318–2326.

[155] Blazquez M, Docherty K, Shennan KI (2001) Association of prohormone convertase 3 with membrane lipid rafts. *J Mol Endocrinol* 27: 107–116.

[156] Arnaoutova I, Smith AM, Coates LC, Sharpe JC, Dhanvantari S, Snell CR, Birch NP, Loh YP (2003) The prohormone processing enzyme PC3 is a lipid raft-associated transmembrane protein. *Biochemistry* 42: 10445–10455.

[157] Lacombe MJ, Mercure C, Dikeakos JD, Reudelhuber TL (2005) Modulation of secretory granule-targeting efficiency by cis and trans compounding of sorting signals. *J Biol Chem* 280: 4803–4807.

[158] Salvas A, Benjannet S, Reudelhuber TL, Chretien M, Seidah NG (2005) Evidence for proprotein convertase activity in the endoplasmic reticulum/early Golgi. *FEBS Lett* 579: 5621–5625.

[159] Stettler H, Suri G, Spiess M (2005) Proprotein convertase PC3 is not a transmembrane protein. *Biochemistry* 44: 5339–5345.

[160] Kirchmair R, Gee P, Hogue-Angeletti R, Laslop A, Fischer-Colbrie R, Winkler H (1992)

Immunological characterization of the endoproteases PC1 and PC2 in adrenal chromaffin granules and in pituitary gland. *FEBS Lett* 297: 302–305.

[161]   Jean F, Basak A, Rondeau N, Benjannet S, Hendy GN, Seidah NG, Chreiten M, Lazure C (1993) Enzymatic characterization of murine and human prohormone convertase-1 (mPC1 and hPC1) expressed in mammalian GH_4C_1 cells. *Biochem J* 292: 891–900.

[162]   Peinado JR, Li H, Johanning K, Lindberg I (2003) Cleavage of recombinant proenkephalin and blockade mutants by prohormone convertases 1 and 2: an in vitro specificity study. *J Neurochem* 87: 868–878.

[163]   Hatsuzawa K, Nagahama M, Takahashi K, Takada K, Murakami K, Nakayama K (1992) Purification and characterization of furin, a Kex2-like processing endoprotease, produced in Chinese hamster ovary cells. *J Biol Chem* 267: 16094–16099.

[164]   Cameron A, Appel J, Houghten RA, Lindberg I (2000) Polyarginines are potent furin inhibitors. *J Biol Chem* 275: 36741–36749.

[165]   Bravo DA, Gleason JB, Sanchez RI, Roth RA, Fuller RS (1994) Accurate and efficient cleavage of the human insulin proreceptor by the human proprotein-processing protease furin. Characterization and kinetic parameters using the purified, secreted soluble protease expressed by a recombinant baculovirus. *J Biol Chem* 269: 25830–25837.

[166]   Lamango NS, Zhu X, Lindberg I (1996) Purification and enzymatic characterization of recombinant prohormone convertase 2: stabilization of activity by 21 kDa 7B2. *Arch Biochem Biophys* 330: 238–250.

[167]   Johanning K, Juliano MA, Juliano L, Lazure C, Lamango NS, Steiner DF, Lindberg I (1998) Specificity of prohormone convertase 2 on proenkephalin and proenkephalin-related substrates. *J Biol Chem* 273: 22672–22680.

[168]   Coates LC, Birch NP (1998) Differential cleavage of provasopressin by the major molecular forms of SPC3. *J Neurochem* 70: 1670–1678.

[169]   Kowalska D, Liu J, Appel JR, Ozawa A, Nefzi A, Mackin RB, Houghten RA, Lindberg I (2009) Synthetic small-molecule prohormone convertase 2 inhibitors. *Mol Pharmacol* 75: 617–625.

[170]   Helwig M, Vivoli M, Fricker LD, Lindberg I (2011) Regulation of neuropeptide processing enzymes by catecholamines in endocrine cells. *Mol Pharmacol* 80: 304–313.

[171]   Jean F, Boudreault A, Basak A, Seidah NG, Lazure C (1995) Fluorescent peptidyl substrates as an aid in studying the substrate specificity of human prohormone convertase PC1 and human furin and designing a potent irreversible inhibitor. *J Biol Chem* 270: 19225–19231.

[172]   Basak A, Jean F, Seidah NG, Lazure C (1994) Design and synthesis of novel inhibitors of prohormone convertases. *Int J Pept Protein Res* 44: 253–261.

[173] Basak A, Schmidt C, Ismail AA, Seidah NG, Chretien M, Lazure C (1995) Peptidyl sub-strates containing unnatural amino acid at the P'1 position are potent inhibitors of prohormone convertases. *Int J Pept Protein Res* 46: 228–237.

[174] Boudreault A, Gauthier D, Lazure C (1998) Proprotein convertase PC1/3-related peptides are potent slow tight- binding inhibitors of murine PC1/3 and Hfurin. *J Biol Chem* 273: 31574–31580.

[175] Basak A, Cooper S, Roberge AG, Banik UK, Chretien M, Seidah NG (1999) Inhibition of proprotein convertases-1, -7 and furin by diterpines of Andrographis paniculata and their succinoyl esters. *Biochem J* 338 (Pt 1): 107–113.

[176] Ozawa A, Peinado JR, Lindberg I (2010) Modulation of prohormone convertase 1/3 prop-erties using site-directed mutagenesis. *Endocrinology* 151: 4437–4445.

[177] Ozawa A, Cai Y, Lindberg I (2007) Production of bioactive peptides in an in vitro system. *Anal Biochem* 366: 182–189.

[178] Rabah N, Gauthier D, Dikeakos JD, Reudelhuber TL, Lazure C (2007) The C-terminal region of the proprotein convertase 1/3 (PC1/3) exerts a bimodal regulation of the enzyme activity in vitro. *FEBS J* 274: 3482–3491.

[179] Bruzzaniti A, Marx R, Mains RE (1999) Activation and routing of membrane-tethered prohormone convertases 1 and 2. *J Biol Chem* 274: 24703–24713.

[180] Bruzzaniti A, Mains RE (2002) Enzymatic activity of soluble and membrane tethered pep-tide pro-hormone convertase 1. *Peptides* 23: 863–875.

[181] Friedman TC, Gordon VM, Leppla SH, Klimpel KR, Birch NP, Loh YP (1995) In vitro processing of anthrax toxin protective antigen by recombinant PC1 (SPC3) and bovine intermediate lobe secretory vesicle membranes. *Arch Biochem Biophys* 316: 5–13.

[182] Nakayama K, Watanabe T, Nakagawa T, Kim WS, Nagahama M, Hosaka M, Hatsuzawa K, Kondoh-Hashiba K, Murakami K (1992) Consensus sequence for precursor processing at mono-arginyl sites. Evidence for the involvement of a Kex2-like endoprotease in precur-sor cleavages at both dibasic and mono-arginyl sites. *J Biol Chem* 267: 16335–16340.

[183] Cameron A, Apletalina EV, Lindberg I (2001) The enzymology of PC1 and PC2. The enzymes XXII: 291–331.

[184] Wardman JH, Zhang X, Gagnon S, Castro LM, Zhu X, Steiner DF, Day R, Fricker LD (2010) Analysis of peptides in prohormone convertase 1/3 null mouse brain using quantita-tive peptidomics. *J Neurochem* 114: 215–225.

[185] Zhang X, Pan H, Peng B, Steiner DF, Pintar JE, Fricker LD (2010) Neuropeptidomic analysis establishes a major role for prohormone convertase-2 in neuropeptide biosynthesis. *J Neurochem* 112: 1168–1179.

[186]   Pan H, Che FY, Peng B, Steiner DF, Pintar JE, Fricker LD (2006) The role of prohormone convertase-2 in hypothalamic neuropeptide processing: a quantitative neuropeptidomic study. *J Neurochem* 98: 1763–1777.

[187]   Pan H, Nanno D, Che FY, Zhu X, Salton SR, Steiner DF, Fricker LD, Devi LA (2005) Neuropeptide processing profile in mice lacking prohormone convertase-1. *Biochemistry* 44: 4939–4948.

[188]   Ozawa A, Lindberg I, Roth B, Kroeze WK (2010) Deorphanization of novel peptides and their receptors. *AAPS J* 12: 378–384.

[189]   Li Q-L, Jansen E, Brent G, Friedman T (2001) Regulation of prohormone convertase 1 (PC1) by thyroid hormone. *Am J Physiol Endocrinol Metab* 280: 160–170.

[190]   Jansen E, Ayoubi TA, Meulemans SM, Van de Ven WJ (1995) Neuroendocrine-specific expression of the human prohormone convertase 1 gene. Hormonal regulation of transcription through distinct cAMP response elements. *J Biol Chem* 270: 15391–15397.

[191]   Jansen E, Ayoubi TA, Meulemans SM, Van de Ven WJ (1997) Cell type-specific protein-DNA interactions at the cAMP response elements of the prohormone convertase 1 promoter. Evidence for additional transactivators distinct from CREB/ATF family members. *J Biol Chem* 272: 2500–2508.

[192]   Hanabusa T, Ohagi S, LaMendola J, Chan SJ, Steiner DF (1994) Nucleotide sequence and analysis of the mouse SPC3 promoter region. *FEBS Lett* 356: 339–341.

[193]   Wen JH, Chen YY, Song SJ, Ding J, Gao Y, Hu QK, Feng RP, Liu YZ, Ren GC, Zhang CY, Hong TP, Gao X, Li LS (2009) Paired box 6 (PAX6) regulates glucose metabolism via proinsulin processing mediated by prohormone convertase 1/3 (PC1/3). *Diabetologia* 52: 504–513.

[194]   Mzhavia N, Qian Y, Feng Y, Che FY, Devi LA, Fricker LD (2002) Processing of proSAAS in neuroendocrine cell lines. *Biochem J* 361: 67–76.

[195]   Jin L, Kulig E, Qian X, Scheithauer BW, Young WF, Jr., Davis DH, Seidah NG, Chretien M, Lloyd RV (1999) Distribution and regulation of proconvertases PC1 and PC2 in human pituitary adenomas. *Pituitary* 1: 187–195.

[196]   Udupi V, Townsend CM, Jr., Greeley GH, Jr. (1998) Stimulation of prohormone convertase-1 mRNA expression by second messenger signaling systems. *Biochem Biophys Res Commun* 246: 463–465.

[197]   Mania-Farnell BL, Botros I, Day R, Davis TP (1996) Differential modulation of prohormone convertase mRNA by second messenger activators in two cholecystokinin-producing cell lines. *Peptides* 17: 47–54.

[198]   Vieau D, Seidah NG, Day R (1995) Mouse insulinoma beta TC3 cells express prodynor-

phin messenger ribonucleic acid and derived peptides: a unique cellular model for the study of prodynorphin biosynthesis and processing. *Endocrinology* 136: 1187–1196.

[199] Birch NP, Hakes DJ, Dixon JE, Mezey E (1994) Distribution and regulation of the candidate prohormone processing enzymes SPC2 and SPC3 in adult rat brain. *Neuropeptides* 27: 307–322.

[200] Sanchez VC, Goldstein J, Stuart RC, Hovanesian V, Huo L, Munzberg H, Friedman TC, Bjorbaek C, Nillni EA (2004) Regulation of hypothalamic prohormone convertases 1 and 2 and effects on processing of prothyrotropin-releasing hormone. *J Clin Invest* 114: 357–369.

[201] Espinosa VP, Ferrini M, Shen X, Lutfy K, Nillni EA, Friedman TC (2007) Cellular colocalization and coregulation between hypothalamic pro-TRH and prohormone convertases in hypothyroidism. *Am J Physiol Endocrinol Metab* 292: E175–186.

[202] Espinosa VP, Liu Y, Ferrini M, Anghel A, Nie Y, Tripathi PV, Porche R, Jansen E, Stuart RC, Nillni EA, Lutfy K, Friedman TC (2008) Differential regulation of prohormone convertase 1/3, prohormone convertase 2 and phosphorylated cyclic-AMP-response element binding protein by short-term and long-term morphine treatment: implications for understanding the "switch" to opiate addiction. *Neuroscience* 156: 788–799.

[203] Marcinkiewicz M, Nagao T, Day R, Seidah NG, Chretien M, Avoli M (1997) Pilocarpine-induced seizures are accompanied by a transient elevation in the messenger RNA expression of the prohormone convertase PC1 in rat hippocampus: comparison with nerve growth factor and brain-derived neurotrophic factor expression. *Neuroscience* 76: 425–439.

[204] Meyer A, Chretien P, Massicotte G, Sargent C, Chretien M, Marcinkiewicz M (1996) Kainic acid increases the expression of the prohormone convertases furin and PC1 in the mouse hippocampus. *Brain Res* 732: 121–132.

[205] Marcinkiewicz M, Savaria D, Marcinkiewicz J (1998) The pro-protein convertase PC1 is induced in the transected sciatic nerve and is present in cultured Schwann cells: comparison with PC5, furin and PC7, implication in pro-BDNF processing. *Brain Res Mol Brain Res* 59: 229–246.

[206] Niquet J, Perez-Martinez L, Guerra M, Grouselle D, Joseph-Bravo P, Charli J (2000) Extracellular matrix proteins increase the expression of pro-TRH and pro-protein convertase PC1 in fetal hypothalamic neurons in vitro. *Brain Res Dev Brain Res* 120: 49–56.

[207] Nilaweera KN, Barrett P, Mercer JG, Morgan PJ (2003) Precursor-protein convertase 1 gene expression in the mouse hypothalamus: differential regulation by ob gene mutation, energy deficit and administration of leptin, and coexpression with prepro-orexin. *Neuroscience* 119: 713–720.

[208] Birch NP, Tracer HL, Hakes DJ, Loh YP (1991) Coordinated regulation of mRNA levels of pro-opiomelanocortin and the candidate processing enzymes PC2 and PC3, but not furin, in rat pituitary intermediate lobe. *Biochem Biophys Res Commun* 179: 1311–1319.

[209] Li Q-L, Jansen E, TC F (1999) Regulation of prohormone convertase 1 (PC1) by gp 130-related cytokines. *Mol Cell Endocrinol* 158: 143–152.

[210] Kobayashi I, Jin L, Ruebel KH, Bayliss JM, Hidehiro O, Lloyd RV (2003) Regulation of cell growth and expression of 7B2, PC2, and PC1/3 by TGFbeta 1 and sodium butyrate in a human pituitary cell line (HP75). *Endocrine* 22: 285–292.

[211] Neerman-Arbez M, Cirulli V, Halban PA (1994) Levels of the conversion endoproteases PC1 (PC3) and PC2 distinguish between insulin-producing pancreatic islet beta cells and non-beta cells. *Biochem J* 300 (Pt 1): 57–61.

[212] Martin S, Caroll R, Benig M, Steiner D (1994) Regulation by glucose of the biosynthesis of PC2, PC3 and pproinsulin in (ob/ob) mouse islets of Langerhans. *FEBS Lett* 356: 279–282.

[213] Alarcon C, Lincolin B, Rhodes C (1994) The biosynthesis of the subtilisin-related propro-tein convertase PC3, but not that of PC2 convertase, is regulated by glucose in parallel to proinsulin biosynthesis in rat pancreatic islets. *J Biol Chem* 268: 4276–4280.

[214] Skelly R, Schuppin G, Ishihara H, Oka Y, Rhodes C (1996) Glucose-regulated transla-tional control of proinsulin biosynthesis with that of the proinsulin endopeptidases PC2 and PC3 in the insulin-producing MIN6 cell line. *Diabetes* 45: 37–43.

[215] Tschuppin G, Rhodes C (1996) Specific co-ordinated regulation of PC3 and PC2 gene expression with that of preproinsulin in insulin-producing beta TC3 cells. *Biochem J* 313: 259–268.

[216] Nie Y, Nakashima M, Brubaker PL, Li QL, Perfetti R, Jansen E, Zambre Y, Pipeleers D, Friedman TC (2000) Regulation of pancreatic PC1 and PC2 associated with increased glucagon-like peptide 1 in diabetic rats. *J Clin Invest* 105: 955–965.

[217] Kilimnik G, Kim A, Steiner DF, Friedman TC, Hara M (2010) Intraislet production of GLP-1 by activation of prohormone convertase 1/3 in pancreatic alpha-cells in mouse models of β-cell regeneration. *Islets* 2: 149–155.

[218] Dhanvantari S, Izzo A, Jansen E, Brubaker PL (2001) Coregulation of glucagon-like peptide-1 synthesis with proglucagon and prohormone convertase 1 gene expression in enteroendocrine GLUTag cells. *Endocrinology* 142: 37–42.

[219] Yu Y, Liu L, Wang X, Liu X, Xie L, Wang G (2010) Modulation of glucagon-like peptide-1 release by berberine: in vivo and in vitro studies. *Biochem Pharmacol* 79: 1000–1006.

[220] Gouraud SS, Heesom K, Yao ST, Qiu J, Paton JF, Murphy D (2007) Dehydration-induced proteome changes in the rat hypothalamo-neurohypophyseal system. *Endocrinology* 148: 3041–3052.

[221] Mihailova A, Karaszewski B, Faergestad EM, Hauser R, Nyka WM, Lundanes E, Greibrokk T (2008) Two-dimensional LC-MS/MS in detection of peptides in hypothalamus of the rat subjected to hypoxic stress. *J Sep Sci* 31: 468–479.

[222] Chakraborty TR, Tkalych O, Nanno D, Garcia AL, Devi LA, Salton SR (2006) Quantification of VGF- and pro-SAAS-derived peptides in endocrine tissues and the brain, and their regulation by diet and cold stress. *Brain Res* 1089: 21–32.

[223] Zhang X, Che FY, Berezniuk I, Sonmez K, Toll L, Fricker LD (2008) Peptidomics of Cpe(fat/fat) mouse brain regions: implications for neuropeptide processing. *J Neurochem* 107: 1596–1613.

[224] Che FY, Yuan Q, Kalinina E, Fricker LD (2005) Peptidomics of Cpe fat/fat mouse hypothalamus: effect of food deprivation and exercise on peptide levels. *J Biol Chem* 280: 4451–4461.

[225] Icimoto MY, Barros NM, Ferreira JC, Marcondes MF, Andrade D, Machado MF, Juliano MA, Judice WA, Juliano L, Oliveira V (2011) Hysteretic behavior of proprotein convertase 1/3 (PC1/3). *PLoS One* 6: e24545.

[226] Berman Y, Mzhavia N, Polonskaia A, Devi LA (2001) Impaired prohormone convertases in cpe^{fat}/cpe^{fat} mice. *J Biol Chem* 276: 1466–1473.

[227] Wolkersdorfer M, Laslop A, Lazure C, Fischer-Colbrie R, Winkler H (1996) Processing of chromogranins in chromaffin cell culture: effects of reserpine and alpha-methyl-p-tyrosine. *Biochem J* 316 (Pt 3): 953–958.

[228] Fricker L, Berman YL, Leiter EH, Devi LA (1996) Carboxypeptidase E activity is deficient in mice with the fat mutation. Effect on peptide processing. *J Biol Chem* 271: 30619–30624.

[229] Kemppainen RJ, Behrend EN (2010) Acute inhibition of carboxypeptidase E expression in AtT-20 cells does not affect regulated secretion of ACTH. *Regul Pept* 165: 174–179.

[230] Wilson SP, Chang KJ, Viveros OH (1980) Synthesis of enkephalins by adrenal medullary chromaffin cells: reserpine increases incorporation of radiolabeled amino acids. *Proc Natl Acad Sci U S A* 77: 4364–4368.

[231] Eiden LE, Zamir N (1986) Metorphamide levels in chromaffin cells increase after treatment with reserpine. *J Neurochem* 46: 1651–1654.

[232] Lindberg I (1986) Reserpine-induced alterations in the processing of proenkephalin in cultured chromaffin cells. Increased amidation. *J Biol Chem* 261: 16317–16322.

[233] Basak A, Koch P, Dupelle M, Fricker LD, Devi LA, Chretien M, Seidah NG (2001) Inhibitory specificity and potency of proSAAS-derived peptides toward proprotein convertase 1. *J Biol Chem* 276: 32720–32728.

[234] Fortenberry Y, Hwang JR, Apletalina EV, Lindberg I (2002) Functional characterization of ProSAAS: similarities and differences with 7B2. *J Biol Chem* 277: 5175–5186.

[235] Fugere M, Day R (2005) Cutting back on pro-protein convertases: the latest approaches to pharmacological inhibition. *Trends Pharmacol Sci* 26: 294–301.

[236] Fugere M, Day R (2002) Inhibitors of the subtilase-like pro-protein convertases (SPCs). *Curr Pharm Des* 8: 549–562.

[237] Garten W, Hallenberger S, Ortmann D, Schafer W, Vey M, Angliker H, Shaw E, Klenk HD (1994) Processing of viral glycoproteins by the subtilisin-like endoprotease furin and its inhibition by specific peptidylchloroalkylketones. *Biochimie* 76: 217–225.

[238] Villemure M, Fournier A, Gauthier D, Rabah N, Wilkes BC, Lazure C (2003) Barley serine proteinase inhibitor 2-derived cyclic peptides as potent and selective inhibitors of convertases PC1/3 and furin. *Biochemistry* 42: 9659–9668.

[239] Becker GL, Sielaff F, Than ME, Lindberg I, Routhier S, Day R, Lu Y, Garten W, Steinmetzer T (2010) Potent inhibitors of furin and furin-like proprotein convertases containing decarboxylated P1 arginine mimetics. *J Med Chem* 53: 1067–1075.

[240] Zhu X, Zhou A, Dey A, Norrbom C, Carroll R, Zhang C, Laurent V, Lindberg I, Ugleholdt R, Holst JJ, Steiner DF (2002) Disruption of PC1/3 expression in mice causes dwarfism and multiple neuroendocrine peptide processing defects. *Proc Natl Acad Sci U S A* 99: 10293–10298.

[241] Mbikay M, Croissandeau G, Sirois F, Anini Y, Mayne J, Seidah NG, Chretien M (2007) A targeted deletion/insertion in the mouse Pcsk1 locus is associated with homozygous embryo preimplantation lethality, mutant allele preferential transmission and heterozygous female susceptibility to dietary fat. *Dev Biol* 306: 584–598.

[242] Lloyd DJ, Bohan S, Gekakis N (2006) Obesity, hyperphagia and increased metabolic efficiency in Pc1 mutant mice. *Hum Mol Genet* 15: 1884–1893.

[243] Zhu X, Orci L, Carroll R, Norrbom C, Ravazzola M, Steiner DF (2002) Severe block in processing of proinsulin to insulin accompanied by elevation of des-64,65 proinsulin intermediates in islets of mice lacking prohormone convertase 1/3. *Proc Natl Acad Sci U S A* 99: 10299–10304.

[244] Ugleholdt R, Zhu X, Deacon CF, Orskov C, Steiner DF, Holst JJ (2004) Impaired intestinal proglucagon processing in mice lacking prohormone convertase 1. *Endocrinology* 145: 1349–1355.

[245] Benzinou M, Creemers JW, Choquet H, Lobbens S, Dina C, *et al.* (2008) Common nonsynonymous variants in PCSK1 confer risk of obesity. *Nat Genet* 40: 943–945.

[246] Wei S, Feng Y, Che FY, Pan H, Mzhavia N, Devi LA, McKinzie AA, Levin N, Richards WG, Fricker LD (2004) Obesity and diabetes in transgenic mice expressing proSAAS. *J Endocrinol* 180: 357–368.

[247] Jackson RS, Creemers JW, Farooqi IS, Raffin-Sanson ML, Varro A, *et al.* (2003) Small-

intestinal dysfunction accompanies the complex endocrinopathy of human proprotein convertase 1 deficiency. *J Clin Invest* 112: 1550–1560.

[248] Farooqi IS, Volders K, Stanhope R, Heuschkel R, White A, Lank E, Keogh J, O'Rahilly S, Creemers JW (2007) Hyperphagia and early-onset obesity due to a novel homozygous missense mutation in prohormone convertase 1/3. *J Clin Endocrinol Metab* 92: 3369–3373.

[249] Creemers JW, Choquet H, Stijnen P, Vatin V, Pigeyre M, *et al.* (2011) Heterozygous mutations causing partial prohormone convertase 1 deficiency contribute to human obesity. Diabetes *In press.*

[250] Gjesing A, Vestmar M, Jorgensen T, Heni M, Holst J, Witte D, Hansen T, Pederson O (2011) The effect of PCSK1 variants on waist, waist-hip ratio and glucose metabolism is modified by sex and glucose tolerance status. *PLoS One* 6.

[251] Kilpelainen TO, Bingham SA, Khaw KT, Wareham NJ, Loos RJ (2009) Association of variants in the PCSK1 gene with obesity in the EPIC-Norfolk study. *Hum Mol Genet* 18: 3496–3501.

[252] Heni M, Haupt A, Schafer SA, Ketterer C, Thamer C, Machicao F, Stefan N, Staiger H, Haring HU, Fritsche A (2010) Association of obesity risk SNPs in PCSK1 with insulin sensitivity and proinsulin conversion. *BMC Med Genet* 11: 86.

[253] Martin LJ, Comuzzie AG, Dupont S, Vionnet N, Dina C, Gallina S, Houari M, Blangero J, Froguel P (2002) A quantitative trait locus influencing type 2 diabetes susceptibility maps to a region on 5q in an extended French family. *Diabetes* 51: 3568–3572.

[254] Rouskas K, Kouvatsi A, Paletas K, Papazoglou D, Tsapas A, Lobbens S, Vatin V, Durand E, Labrune Y, Delplanque J, Meyre D, Froguel P (2011) Common Variants in FTO, MC4R, TMEM18, PRL, AIF1, and PCSK1 Show Evidence of Association With Adult Obesity in the Greek Population. *Obesity* (Silver Spring).

[255] Lin E, Pei D, Huang YJ, Hsieh CH, Wu LS (2009) Gene-gene interactions among genetic variants from obesity candidate genes for nonobese and obese populations in type 2 diabetes. *Genet Test Mol Biomarkers* 13: 485–493.

[256] Goossens GH, Petersen L, Blaak EE, Hul G, Arner P, Astrup A, Froguel P, Patel K, Pedersen O, Polak J, Oppert JM, Martinez JA, Sorensen TI, Saris WH (2009) Several obesity- and nutrient-related gene polymorphisms but not FTO and UCP variants modulate postabsorptive resting energy expenditure and fat-induced thermogenesis in obese individuals: the NUGENOB study. *Int J Obes* (Lond) 33: 669–679.

[257] Mbikay M, Sirois F, Nkongolo KK, Basak A, Chretien M (2011) Effects of rs6234/rs6235 and rs6232/rs6234/rs6235 PCSK1 single-nucleotide polymorphism clusters on proprotein convertase 1/3 biosynthesis and activity. *Mol Genet Metab* 104: 682–687.

[258] Chagnon YC, Rice T, Perusse L, Borecki IB, Ho-Kim MA, Lacaille M, Pare C, Bouchard

L, Gagnon J, Leon AS, Skinner JS, Wilmore JH, Rao DC, Bouchard C (2001) Genomic scan for genes affecting body composition before and after training in Caucasians from HERITAGE. *J Appl Physiol* 90: 1777–1787.

[259]   Hager J, Dina C, Francke S, Dubois S, Houari M, Vatin V, Vaillant E, Lorentz N, Basdevant A, Clement K, Guy-Grand B, Froguel P (1998) A genome-wide scan for human obesity genes reveals a major susceptibility locus on chromosome 10. *Nat Genet* 20: 304–308.

[260]   Chen G, Adeyemo AA, Johnson T, Zhou J, Amoah A, *et al.* (2005) A genome-wide scan for quantitative trait loci linked to obesity phenotypes among West Africans. *Int J Obes* (Lond) 29: 255–259.

[261]   Bell CG, Benzinou M, Siddiq A, Lecoeur C, Dina C, Lemainque A, Clement K, Basdevant A, Guy-Grand B, Mein CA, Meyre D, Froguel P (2004) Genome-wide linkage analysis for severe obesity in french caucasians finds significant susceptibility locus on chromosome 19q. *Diabetes* 53: 1857–1865.

[262]   Chang YC, Chiu YF, Shih KC, Lin MW, Sheu WH, Donlon T, Curb JD, Jou YS, Chang TJ, Li HY, Chuang LM (2010) Common PCSK1 haplotypes are associated with obesity in the Chinese population. *Obesity* (Silver Spring) 18: 1404–1409.

[263]   Sun J, Zhang C, Fang X, Lei C, Lan X, Chen H (2010) Novel single nucleotide polymorphisms of the caprine PC1 gene and association with growth traits. *Biochem Genet* 48: 779–788.

[264]   Wideman RD, Gray SL, Covey SD, Webb GC, Kieffer TJ (2009) Transplantation of PC1/3-Expressing alpha-cells improves glucose handling and cold tolerance in leptin-resistant mice. *Mol Ther* 17: 191–198.

[265]   Tatsumi KI, Tanaka S, Takano T, Tahara S, Murakami Y, Takao T, Hashimoto K, Kato Y, Teramoto A, Amino N (2003) Frequent appearance of autoantibodies against prohormone convertase 1/3 and neuroendocrine protein 7B2 in patients with nonfunctioning pituitary macroadenoma. *Endocrine* 22: 335–340.

[266]   Tzimas GN, Chevet E, Jenna S, Nguyen DT, Khatib AM, Marcus V, Zhang Y, Chretien M, Seidah N, Metrakos P (2005) Abnormal expression and processing of the proprotein convertases PC1 and PC2 in human colorectal liver metastases. *BMC Cancer* 5: 149.

[267]   Blanchard A, Iwasiow B, Yarmill A, Fresnosa A, Silha J, Myal Y, Murphy LC, Chretien M, Seidah N, Shiu RP (2009) Targeted production of proprotein convertase PC1 enhances mammary development and tumorigenesis in transgenic mice. *Can J Physiol Pharmacol* 87: 831–838.

[268]   Park JW, Ji YI, Choi YH, Kang MY, Jung E, Cho SY, Cho HY, Kang BK, Joung YS, Kim DH, Park SC, Park J (2009) Candidate gene polymorphisms for diabetes mellitus, cardiovascular disease and cancer are associated with longevity in Koreans. *Exp Mol Med* 41: 772–781.

[269] Davidsson P, Sjogren M, Andreasen N, Lindbjer M, Nilsson CL, Westman-Brinkmalm A, Blennow K (2002) Studies of the pathophysiological mechanisms in frontotemporal dementia by proteome analysis of CSF proteins. *Brain Res Mol Brain Res* 109: 128–133.

[270] Bayes A, Grant SG (2009) Neuroproteomics: understanding the molecular organization and complexity of the brain. *Nat Rev Neurosci* 10: 635–646.

[271] Siekhaus DE, Fuller RS (1999) A role for amontillado, the Drosophila homolog of the neuropeptide precursor processing protease PC2, in triggering hatching behavior. *J Neurosci* 19: 6942–6954.

[272] Lindberg I, Tu B, Muller L, Dickerson IM (1998) Cloning and functional analysis of C. elegans 7B2. *DNA Cell Biol* 17: 727–734.

[273] Kovaleva ES, Yakovlev AG, Masler EP, Chitwood DJ (2002) Human proprotein convertase 2 homologue from a plant nematode: cloning, characterization, and comparison with other species. *FASEB J* 16: 1099–1101.

[274] Nagle GT, Garcia AT, Knock SL, Gorham EL, Van Heumen WR, Kurosky A (1995) Molecular cloning, cDNA sequence, and localization of a prohormone convertase (PC2) from the Aplysia atrial gland. *DNA Cell Biol* 14: 145–154.

[275] Srivastava M, Simakov O, Chapman J, Fahey B, Gauthier ME, *et al.* (2010) The Amphimedon queenslandica genome and the evolution of animal complexity. *Nature* 466: 720–726.

[276] Bertrand S, Camasses A, Paris M, Holland ND, Escriva H (2006) Phylogenetic analysis of Amphioxus genes of the proprotein convertase family, including aPC6C, a marker of epithelial fusions during embryology. *Int J Biol Sci* 2: 125–132.

[277] Creemers JW, Usac EF, Bright NA, Van de Loo JW, Jansen E, Van de Ven WJ, Hutton JC (1996) Identification of a transferable sorting domain for the regulated pathway in the prohormone convertase PC2. *J Biol Chem* 271: 25284–25291.

[278] Taylor NA, Shennan KI, Cutler DF, Docherty K (1997) Mutations within the propeptide, the primary cleavage site or the catalytic site, or deletion of C-terminal sequences, prevents secretion of proPC2 from transfected COS-7 cells. *Biochem J* 321 (Pt 2): 367–373.

[279] Shennan KIJ, Taylor NA, Docherty K (1994) Calcium- and pH-dependent aggregation and membrane association of the precursor of the prohormone convertase PC2. *J Biol Chem* 269: 18646–18650.

[280] Blazquez M, Thiele C, Huttner WB, Docherty K, Shennan KI (2000) Involvement of the membrane lipid bilayer in sorting prohormone convertase 2 into the regulated secretory pathway. *Biochem J* 349: 843–852.

[281] Benjannet S, Lusson J, Savaria D, Chretien M, Seidah NG (1995) Structure-function studies on the biosynthesis and bioactivity of the precursor convertase PC2 and the formation of the PC2/7B2 complex. *FEBS Lett* 362: 151–155.

[282] Zhu X, Muller L, Mains RE, Lindberg I (1998) Structural elements of PC2 required for interaction with its helper protein 7B2. *J Biol Chem* 273: 1158–1164.

[283] Taylor NA, Jan G, Scougall KT, Docherty K, Shennan KI (1998) Sorting of PC2 to the regulated secretory pathway in AtT20 cells. *J Mol Endocrinol* 21: 209–216.

[284] Assadi M, Sharpe JC, Snell C, Loh YP (2004) The C-terminus of prohormone convertase 2 is sufficient and necessary for Raft association and sorting to the regulated secretory pathway. *Biochemistry* 43: 7798–7807.

[285] Henrich S, Lindberg I, Bode W, Than ME (2005) Proprotein convertase models based on the crystal structures of furin and kexin: explanation of their specificity. *J Mol Biol* 345: 211–227.

[286] Than ME, Henrich S, Bourenkov GP, Bartunik HD, Huber R, Bode W (2005) The endoproteinase furin contains two essential Ca2+ ions stabilizing its N-terminus and the unique S1 specificity pocket. *Acta Crystallogr D Biol Crystallogr* 61: 505–512.

[287] Braks JAM, Martens GJM (1994) 7B2 is a neuroendocrine chaperone that transiently interacts with prohormone convertase PC2 in the secretory pathway. *Cell* 78: 263–273.

[288] Benjannet S, Savaria D, Chretien M, Seidah NG (1995) 7B2 is a specific intracellular binding protein of the prohormone convertase PC2. *J Neurochem* 64: 2303–2311.

[289] Zhu X, Lindberg I (1995) 7B2 facilitates the maturation of proPC2 in neuroendocrine cells and is required for the expression of enzymatic activity. *J Cell Biol* 129: 1641–1650.

[290] Seidel B, Dong W, Savaria D, Zheng M, Pintar JE, Day R (1998) Neuroendocrine protein 7B2 is essential for proteolytic conversion and activation of proprotein convertase 2 in vivo. *DNA Cell Biol* 17: 1017–1029.

[291] Westphal CH, Muller L, Zhou A, Zhu X, Bonner-Weir S, Schambelan M, Steiner DF, Lindberg I, Leder P (1999) The neuroendocrine protein 7B2 is required for peptide hormone processing in vivo and provides a novel mechanism for pituitary Cushing's disease. *Cell* 96: 689–700.

[292] Paquet L, Bergeron F, Boudreault A, Seidah NG, Chretien M, Mbikay M, Lazure C (1994) The neuroendocrine precursor 7B2 is a sulfated protein proteolytically processed by a ubiquitous furin-like convertase. *J Biol Chem* 269: 19279–19285.

[293] Iguchi H, Chan JSD, Seidah NG, Chretien M (1984) Tissue distribution and molecular forms of a novel pituitary protein in the rat. *Neuroendocrinology* 39: 453–458.

[294] Suzuki H, Ghatei MA, Williams SJ, Uttenthal LO, Facer P, Bishop AE, Polak JM, Bloom SR (1986) Production of pituitary protein 7B2 immunoreactivity by endocrine tumors and its possible diagnostic value. *J Clin Endocrinol Metab* 63: 758–765.

[295] Ayoubi TY, van Duijnhoven HLP, van de Ven WJM, Jenks BG, Roubos EW, Martens GJM (1990) The neuroendocrine polypeptide 7B2 is a precursor protein. *J Biol Chem* 265: 15644–15647.

[296]   Martens GJ, Braks JA, Eib DW, Zhou Y, Lindberg I (1994) The neuroendocrine polypeptide 7B2 is an endogenous inhibitor of prohormone convertase PC2. *Proc Natl Acad Sci U S A* 91: 5784–5787.

[297]   Apletalina EV, Juliano MA, Juliano L, Lindberg I (2000) Structure-function analysis of the 7B2 CT peptide. *Biochem Biophys Res Commun* 267: 940–942.

[298]   van Horssen AM, Martens GJ (1998) Mapping of the domain in the neuroendocrine protein 7B2 important for its helper function towards prohormone convertase PC2. *Mol Cell Endocrinol* 137: 7–12.

[299]   Zhu X, Rouille Y, Lamango NS, Steiner DF, Lindberg I (1996) Internal cleavage of the inhibitory 7B2 carboxyl-terminal peptide by PC2: a potential mechanism for its inactivation. *Proc Natl Acad Sci U S A* 93: 4919–4924.

[300]   Hwang JR, Siekhaus DE, Fuller RS, Taghert PH, Lindberg I (2000) Interaction of Drosophila melanogaster prohormone convertase 2 and 7B2. Insect cell-specific processing and secretion. *J Biol Chem* 275: 17886–17893.

[301]   Martens GJM (1988) Cloning and sequence analysis of human pituitary cDNA encoding the novel polypeptide 7B2. *FEBS Lett* 234: 160–164.

[302]   Mehrabian M, Wen PZ, Fisler J, Davis RC, Lusis AJ (1998) Genetic loci controlling body fat, lipoprotein metabolism, and insulin levels in a multifactorial mouse model. *J Clin Invest* 101: 2485–2496.

[303]   Paquet L, Massie B, Mains RE (1996) Proneuropeptide Y processing in large dense-core vesicles: manipulation of prohormone convertase expression in sympathetic neurons using adenoviruses. *J Neurosci* 16: 964–973.

[304]   Tang SS, Zhang JH, Liu HX, Li HZ (2009) PC2/CPE-mediated pro-protein processing in tumor cells and its differentiated cells or tissues. *Mol Cell Endocrinol* 303: 43–49.

[305]   Lansac G, Dong W, Dubois CM, Benlarbi N, Afonso C, Fournier I, Salzet M, Day R (2006) Lipopolysaccharide mediated regulation of neuroendocrine associated proprotein convertases and neuropeptide precursor processing in the rat spleen. *J Neuroimmunol* 171: 57–71.

[306]   Schiller M, Raghunath M, Kubitscheck U, Scholzen TE, Fisbeck T, Metze D, Luger TA, Bohm M (2001) Human dermal fibroblasts express prohormone convertases 1 and 2 and produce proopiomelanocortin-derived peptides. *J Invest Dermatol* 117: 227–235.

[307]   Beaubien G, Schafer MK, Weihe E, Dong W, Chretien M, Seidah NG, Day R (1995) The distinct gene expression of the pro-hormone convertases in the rat heart suggests potential substrates. *Cell Tissue Res* 279: 539–549.

[308]   Guest PC, Arden SD, Bennett DL, Clark A, Rutherford NG, Hutton JC (1992) The post-translational processing and intracellular sorting of PC2 in the islets of Langerhans. *J Biol Chem* 267: 22401–22406.

[309]   Shen FS, Seidah NG, Lindberg I (1993) Biosynthesis of the prohormone convertase PC2 in Chinese hamster ovary cells and in rat insulinoma cells. *J Biol Chem* 268: 24910–24915.

[310]   Creemers JW, van de Loo JW, Plets E, Hendershot LM, Van De Ven WJ (2000) Binding of BiP to the processing enzyme lymphoma proprotein convertase prevents aggregation, but slows down maturation. *J Biol Chem* 275: 38842–38847.

[311]   Uehara M, Yaoi Y, Suzuki M, Takata K, Tanaka S (2001) Differential localization of prohormone convertases PC1 and PC2 in two distinct types of secretory granules in rat pituitary gonadotrophs. *Cell Tissue Res* 304: 43–49.

[312]   Waldbieser GC, Aimi J, Dixon JE (1991) Cloning and characterization of the rat complementary deoxyribonucleic acid and gene encoding the neuroendocrine peptide 7B2. *Endocrinology* 128: 3228–3236.

[313]   D'Anjou F, Bergeron LJ, Larbi NB, Fournier I, Salzet M, Perreault JP, Day R (2004) Silencing of SPC2 expression using an engineered delta ribozyme in the mouse betaTC-3 endocrine cell line. *J Biol Chem* 279: 14232–14239.

[314]   Ozawa A, Lick AN, Lindberg I (2011) Processing of proaugurin is required to suppress proliferation of tumor cell lines. *Mol Endocrinol* 25: 776–784.

[315]   Marcinkiewicz M, Ramla D, Seidah NG, Chretien M (1994) Developmental expression of the prohormone convertases PC1 and PC2 in mouse pancreatic islets. *Endocrinology* 135: 1651–1660.

[316]   Holling TM, van Herp F, Durston AJ, Martens GJ (2000) Differential onset of expression of mRNAs encoding proopiomelanocortin, prohormone convertases 1 and 2, and granin family members during Xenopus laevis development. *Brain Res Mol Brain Res* 75: 70–75.

[317]   Lee SN, Kacprzak MM, Day R, Lindberg I (2007) Processing and trafficking of a prohormone convertase 2 active site mutant. *Biochem Biophys Res Commun* 355: 825–829.

[318]   Muller L, Zhu X, Lindberg I (1997) Mechanism of the facilitation of PC2 maturation by 7B2: involvement in ProPC2 transport and activation but not folding. *J Cell Biol* 139: 625–638.

[319]   Lamango NS, Apletalina E, Liu J, Lindberg I (1999) The proteolytic maturation of prohormone convertase 2 (PC2) is a pH- driven process. *Arch Biochem Biophys* 362: 275–282.

[320]   Bailyes EM, Shennan KIJ, Usac EF, Arden SD, Guest PC, Docherty K, Hutton JC (1995) Differences between the catalytic properties of recombinant human PC2 and endogenous rat PC2. *Biochem J* 309: 587–594.

[321]   Xu H, Shields D (1994) Prosomatostatin processing in permeabilized cells. Endoproteolytic cleavage is mediated by a vacuolar ATPase that generates an acidic pH in the trans-Golgi network. *J Biol Chem* 269: 22875–22881.

[322]   Jansen EJ, Hafmans TG, Martens GJ (2010) V-ATPase-mediated granular acidification

is regulated by the V-ATPase accessory subunit Ac45 in POMC-producing cells. *Mol Biol Cell* 21: 3330–3339.

[323]   Shennan KIJ, Smeekens SP, Steiner DF, Docherty K (1991) Characterization of PC2, a mammalian Kex2 homologue, following expression of the cDNA in microinjection Xenopus oocytes. *FEBS Lett* 284: 277–280.

[324]   Lee SN, Lindberg I (2008) 7B2 prevents unfolding and aggregation of prohormone convertase 2. *Endocrinology* 149: 4116–4127.

[325]   Braks JAM, Van Horssen AM, Martens GJM (1996) Dissociation of the complex between the neuroendocrine chaperone 7B2 and prohormone convertase PC2 is not associated with proPC2 maturation. *Eur J Biochem* 238: 505–510.

[326]   Lee SN, Hwang JR, Lindberg I (2006) Neuroendocrine protein 7B2 can be inactivated by phosphorylation within the secretory pathway. *J Biol Chem* 281: 3312–3320.

[327]   Zhu X, Lamango NS, Lindberg I (1996) Involvement of a polyproline helix-like structure in the interaction of 7B2 with prohormone convertase 2. *J Biol Chem* 271: 23582–23587.

[328]   Muller L, Zhu P, Juliano MA, Juliano L, Lindberg I (1999) A 36-residue peptide contains all of the information required for 7B2-mediated activation of prohormone convertase 2. *J Biol Chem* 274: 21471–21477.

[329]   Hwang JR, Lindberg I (2001) Inactivation of the 7B2 inhibitory CT peptide depends on a functional furin cleavage site. *J Neurochem* 79: 437–444.

[330]   Chaudhuri B, Stephen C, Huijbregts R, Martens G (1995) The neuroendocrine protein 7B2 acts as a molecular chaperone in the in vitro folding of human insulin-like growth factor-1 secreted from yeast. *Biochem Biophys Res Commun* 211: 417–425.

[331]   Benjannet S, Mamarbachi AM, Hamelin J, Savaria D, Munzer JS, Chretien M, Seidah NS (1998) Residues unique to the prohormone convertase PC2 modulate its autoactivation, binding to 7B2, and enzymatic activity. *FEBS Lett* 428: 37–42.

[332]   Apletalina EV, Muller L, Lindberg I (2000) Mutations in the catalytic domain of prohormone convertase 2 result in decreased binding to 7B2 and loss of inhibition with 7B2 C-terminal peptide. *J Biol Chem* 275: 14667–14677.

[333]   Rockwell NC, Krysan DJ, Fuller RS (2000) Synthesis of peptidyl methylcoumarin esters as substrates and active-site titrants for the prohormone processing proteases Kex2 and PC2. *Anal Biochem* 280: 201–208.

[334]   Shennan KIJ, Seal AJ, Smeekens SP, F. SD, Docherty K (1991) Site directed mutagenesis and expression of PC2 in microinjected Xenopus oocytes. *J Biol Chem* 266: 24011–24017.

[335]   Azaryan AV, Krieger TJ, Hook VY (1995) Purification and characteristics of the candidate prohormone processing proteases PC2 and PC1/3 from bovine adrenal medulla chromaffin granules. *J Biol Chem* 270: 8201–8208.

[336] Fahnestock M, Zhu W (1999) Expression of human prohormone convertase PC2 in a baculovirus-insect cell system. *DNA Cell Biol* 18: 409–417.

[337] Braks JAM, Martens GJM (1995) The neuroendocrine chaperone 7B2 can enhance in vitro POMC cleavage by prohormone convertase PC2. *FEBS Lett* 371: 154–158.

[338] Lindberg I, van den Hurk WH, Bui C, Batie CJ (1995) Enzymatic characterization of immunopurified prohormone convertase 2: potent inhibition by a 7B2 peptide fragment. *Biochemistry* 34: 5486–5493.

[339] Li QL, Naqvi S, Shen X, Liu YJ, Lindberg I, Friedman TC (2003) Prohormone convertase 2 enzymatic activity and its regulation in neuro-endocrine cells and tissues. *Regul Pept* 110: 197–205.

[340] Berman Y, Mzhavia N, Polonskaia A, Furuta M, Steiner DF, Pintar JE, Devi LA (2000) Defective prodynorphin processing in mice lacking prohormone convertase PC2. *J Neurochem* 75: 1763–1770.

[341] Thomas L, Leduc R, Thorne BA, Smeekens SP, Steiner D, Thomas G (1991) Kex2-like endoproteases PC2 and PC3 accurately cleave a model prohormone in mammalian cells: evidence for a common core of neuroendocrine processing enzymes. *Proc Natl Acad Sci U S A* 88: 5297–5301.

[342] Remacle AG, Shiryaev SA, Oh ES, Cieplak P, Srinivasan A, Wei G, Liddington RC, Ratnikov BI, Parent A, Desjardins R, Day R, Smith JW, Lebl M, Strongin AY (2008) Substrate cleavage analysis of furin and related proprotein convertases. A comparative study. *J Biol Chem* 283: 20897–20906.

[343] Day R, Lazure C, Basak A, Boudreault A, Limperis P, Dong W, Lindberg I (1998) Prodynorphin processing by proprotein convertase 2. Cleavage at single basic residues and enhanced processing in the presence of carboxypeptidase activity. *J Biol Chem* 273: 829–836.

[344] Mathis JP, Lindberg I (1992) Posttranslational processing of proenkephalin in AtT-20 cells: evidence for cleavage at a Lys-Lys site. *Endocrinology* 131: 2287–2296.

[345] Kacprzak MM, Than ME, Juliano L, Juliano MA, Bode W, Lindberg I (2005) Mutations of the PC2 substrate binding pocket alter enzyme specificity. *J Biol Chem* 280: 31850–31858.

[346] Li QL, Jansen E, Brent GA, Naqvi S, Wilber JF, Friedman TC (2000) Interactions between the prohormone convertase 2 promoter and the thyroid hormone receptor. *Endocrinology* 141: 3256–3266.

[347] Katz LS, Gosmain Y, Marthinet E, Philippe J (2009) Pax6 regulates the proglucagon processing enzyme PC2 and its chaperone 7B2. *Mol Cell Biol* 29: 2322–2334.

[348] Jansen E, Ayoubi TA, Meulemans SM, Van De Ven WJ (1997) Regulation of human prohormone convertase 2 promoter activity by the transcription factor EGR-1. *Biochem J* 328 (Pt 1): 69–74.

[349]   Mbikay M, Raffin-Sanson ML, Sirois F, Kalenga L, Chretien M, Seidah NG (2002) Characterization of a repressor element in the promoter region of proprotein convertase 2 (PC2) gene. *Brain Res Mol Brain Res* 102: 35–47.

[350]   D'Alessandro R, Klajn A, Stucchi L, Podini P, Malosio ML, Meldolesi J (2008) Expression of the neurosecretory process in PC12 cells is governed by REST. *J Neurochem* 105: 1369–1383.

[351]   Jing E, Nillni EA, Sanchez VC, Stuart RC, Good DJ (2004) Deletion of the Nhlh2 transcription factor decreases the levels of the anorexigenic peptides alpha melanocyte-stimulating hormone and thyrotropin-releasing hormone and implicates prohormone convertases I and II in obesity. *Endocrinology* 145: 1503–1513.

[352]   Oyarce AM, Hand TA, Mains RE, Eipper BA (1996) Dopaminergic regulation of secretory granule-associated proteins in rat intermediate pituitary. *J Neurochem* 67: 229–241.

[353]   Saiardi A, Borelli E (1998) Absence of dopaminergic control on melanotrophs leads to Cushing`s-like syndrome in mice. *Mol Endocrinol* 12: 1133–1139.

[354]   Bhat RV, Tausk FA, Baraban JM, Mains RE, Eipper BA (1993) Rapid increases in peptide processing enzyme expression in hippocampal neurons. *J Neurochem* 61: 1315–1322.

[355]   Van Horssen AM, Van den Hurk WH, Bailyes EM, Hutton JC, Martens GJM, Lindberg I (1995) Identification of the region within the neuroendocrine polypeptide 7B2 responsible for the inhibition of prohormone convertase PC2. *J Biol Chem* 270: 14292–14296.

[356]   Fortenberry Y, Liu J, Lindberg I (1999) The role of the 7B2 CT peptide in the inhibition of prohormone convertase 2 in endocrine cell lines. *J Neurochem* 73: 994–1003.

[357]   Palmer DJ, Christie DL (1992) Identification of molecular aggregates containing glycoproteins III, J, K(carboxypeptidase H), and H (kex2-related proteins) in the soluble and membrane fractions of adrenal medullary chromaffin granules. *J Biol Chem* 267: 19806–19812.

[358]   Basak A, Gauthier D, N.G. S, Lazure C (1997) Synthetic proregion-related peptides of proprotein convertases, PC1 and furin, represent potent inhibitors of each protease. In: Peptides: Frontiers of Peptide Science N, TE, USA., editor. Proceedings of the XVth American Peptide Symposium. Dordrecht, The Netherlands: Kluwer Academic Publishers. pp. 676–677.

[359]   Feliciangeli SF, Thomas L, Scott GK, Subbian E, Hung CH, Molloy SS, Jean F, Shinde U, Thomas G (2006) Identification of a pH sensor in the furin propeptide that regulates enzyme activation. *J Biol Chem* 281: 16108–16116.

[360]   Cornwall GA, Cameron A, Lindberg I, Hardy DM, Cormier N, Hsia N (2003) The cystatin-related epididymal spermatogenic protein inhibits the serine protease prohormone convertase 2. *Endocrinology* 144: 901–908.

[361]   Braks JA, Guldemond KC, van Riel MC, Coenen AJ, Martens GJ (1992) Structure and expression of Xenopus prohormone convertase PC2. *FEBS Lett* 305: 45–50.

[362]   Barbero P, Kitabgi P (1999) Protein 7B2 is essential for the targeting and activation of PC2 into the regulated secretory pathway of rMTC 6-23 cells. *Biochem Biophys Res Commun* 257: 473–479.

[363]   Dong W, Day R (2002) Gene expression of proprotein convertases in individual rat anterior pituitary cells and their regulation in corticotrophs mediated by glucocorticoids. *Endocrinology* 143: 254–262.

[364]   Petit-Turcotte C, Paquin J (2000) Cordinate regulation of neuroendocrine convertase PC2 and peptide 7B2 in P19 neurons. *Peptides* 21: 365–372.

[365]   Jeannotte R, Paquin J, Petit-Turcotte C, Day R (1997) Convertase PC2 and the neuroendocrine polypeptide 7B2 are co-induced and processed during neuronal differentiation of P19 embryonal carcinoma cells. *DNA Cell Biol* 16: 1175–1187.

[366]   Schmidt G, Sirois F, Anini Y, Kauri LM, Gyamera-Acheampong C, Fleck E, Scott FW, Chretien M, Mbikay M (2006) Differences of pancreatic expression of 7B2 between C57BL/6J and C3H/HeJ mice and genetic polymorphisms at its locus (Sgne1). *Diabetes* 55: 452–459.

[367]   Helwig M, Lee SN, Hwang JR, Ozawa A, Medrano JF, Lindberg I (2011) Dynamic modulation of PC2-mediated precursor processing by 7B2: preferential effect on glucagon synthesis. *J Biol Chem 286*: 42504–42513.

[368]   Farber CR, Chitwood J, Lee SN, Verdugo RA, Islas-Trejo A, Rincon G, Lindberg I, Medrano JF (2008) Overexpression of Scg5 increases enzymatic activity of PCSK2 and is inversely correlated with body weight in congenic mice. *BMC Genet* 9: 34.

[369]   Yuan B, Meudt J, Blank R, Feng J, Drezner M (2010) Hexa-D-arginine reversal of osteoblast 7B2 dysregulation in Hyp mice Normalizes the HYP Biochemical Phenotype. ASBMR Abstracts.

[370]   Braks JA, Broers CA, Danger JM, Martens GJ (1996) Structural organization of the gene encoding the neuroendocrine chaperone 7B2. *Eur J Biochem* 236: 60–67.

[371]   Tadros H, Schmidt G, Sirois F, Mbikay M (2011) Regulation of 7B2 mRNA Translation: Dissecting the Role of Its 5'-Untranslated Region. *Methods Mol Biol* 768: 217–230.

[372]   Waha A, Felsberg J, Hartmann W, Hammes J, Knesebeck AV, Endl E, Pietsch T (2011) Frequent epigenetic inactivation of the chaperone SGNE1/7B2 in human gliomas. *Int J Cancer* In press.

[373]   Waha A, Koch A, Hartmann W, Milde U, Felsberg J, Hubner A, Mikeska T, Goodyer CG, Sorensen N, Lindberg I, Wiestler OD, Pietsch T (2007) SGNE1/7B2 is epigenetically altered and transcriptionally downregulated in human medulloblastomas. *Oncogene* 26: 5662–5668.

[374] Furuta M, Carroll R, Martin S, Swift HH, Ravazzola M, Orci L, Steiner DF (1998) Incomplete processing of proinsulin to insulin accompanied by elevation of Des-31,32 proinsulin intermediates in islets of mice lacking active PC2. *J Biol Chem* 273: 1–7.

[375] Furuta M, Yano H, Zhou A, Rouille Y, Holst JJ, Carroll R, Ravazzola M, Orci L, Furuta H, Steiner DF (1997) Defective prohormone processing and altered pancreatic islet morphology in mice lacking active SPC2. *Proc Natl AcadSci U S A* 94: 6646–6651.

[376] Vincent M, Guz Y, Rozenberg M, Webb G, Furuta M, Steiner D, Teitelman G (2003) Abrogation of protein convertase 2 activity results in delayed islet cell differentiation and maturation, increased alpha-cell proliferation, and islet neogenesis. *Endocrinology* 144: 4061–4069.

[377] Furuta M, Zhou A, Webb G, Carroll R, Ravazzola M, Orci L, Steiner DF (2001) Severe defect in proglucagon processing in islet A-cells of prohormone convertase 2 null mice. *J Biol Chem* 276: 27197–27202.

[378] Webb GC, Akbar MS, Zhao C, Swift HH, Steiner DF (2002) Glucagon replacement via micro-osmotic pump corrects hypoglycemia and alpha-cell hyperplasia in prohormone convertase 2 knockout mice. *Diabetes* 51: 398–405.

[379] Croissandeau G, Wahnon F, Yashpal K, Seidah NG, Coderre TJ, Chretien M, Mbikay M (2006) Increased stress-induced analgesia in mice lacking the proneuropeptide convertase PC2. *Neurosci Lett* 406: 71–75.

[380] Sarac MS, Zieske AW, Lindberg I (2002) The lethal form of Cushing's in 7B2 null mice is caused by multiple metabolic and hormonal abnormalities. *Endocrinology* 143: 2324–2332.

[381] Laurent V, Kimble A, Peng B, Zhu P, Pintar JE, Steiner DF, Lindberg I (2002) Mortality in 7B2 null mice can be rescued by adrenalectomy: involvement of dopamine in ACTH hypersecretion. *Proc Natl Acad Sci U S A* 99: 3087–3092.

[382] Rouille Y, Martin S, Steiner DF (1995) Differential processing of proglucagon by the subtilisin-like prohormone convertases PC2 and PC3 to generate either glucagon or glucagon-like peptide. *J Biol Chem* 270: 26488–26496.

[383] Laurent V, Jaubert-Miazza L, Desjardins R, Day R, Lindberg I (2004) Biosynthesis of proopiomelanocortin-derived peptides in prohormone convertase 2 and 7B2 null mice. *Endocrinol* 145: 519–528.

[384] Peinado JR, Laurent V, Lee SN, Peng BW, Pintar JE, Steiner DF, Lindberg I (2005) Strain-dependent influences on the hypothalamo-pituitary-adrenal axis profoundly affect the 7B2 and PC2 null phenotypes. *Endocrinology* 146: 3438–3444.

[385] Lee SN, Peng B, Desjardins R, Pintar JE, Day R, Lindberg I (2007) Strain-specific steroidal control of pituitary function. *J Endocrinol* 192: 515–525.

[386] Rehfeld JF, Lindberg I, Friis-Hansen L (2002) Increased synthesis but decreased processing

of neuronal proCCK in prohormone convertase 2 and 7B2 knockout animals. *J Neurochem* 83: 1329–1337.

[387] Rehfeld JF, Lindberg I, Friis-Hansen L (2002) Progastrin processing differs in 7B2 and PC2 knockout animals: a role for 7B2 independent of action on PC2. *FEBS Lett* 510: 89–93.

[388] Rayburn LY, Gooding HC, Choksi SP, Maloney D, Kidd AR, 3rd, Siekhaus DE, Bender M (2003) amontillado, the Drosophila homolog of the prohormone processing protease PC2, is required during embryogenesis and early larval development. *Genetics* 163: 227–237.

[389] Reiher W, Shirras C, Kahnt J, Baumeister S, Isaac RE, Wegener C (2011) Peptidomics and peptide hormone processing in the Drosophila midgut. *J Proteome Res* 10: 1881–1892.

[390] Rhea JM, Wegener C, Bender M (2010) The proprotein convertase encoded by amontillado (amon) is required in Drosophila corpora cardiaca endocrine cells producing the glucose regulatory hormone AKH. *PLoS Genet* 6: e1000967.

[391] Wegener C, Herbert H, Kahnt J, Bender M, Rhea JM (2011) Deficiency of prohormone convertase dPC2 (AMONTILLADO) results in impaired production of bioactive neuropeptide hormones in Drosophila. *J Neurochem* 118: 581–595.

[392] Kass J, Jacob TC, Kim P, Kaplan JM (2001) The EGL-3 proprotein convertase regulates mechanosensory responses of Caenorhabditis elegans. *J Neurosci* 21: 9265–9272.

[393] He KW, Shen LL, Zhou WW, Wang DY (2009) Regulation of aging by unc-13 and sbt-1 in Caenorhabditis elegans is temperature-dependent. *Neurosci Bull* 25: 335–342.

[394] Gomez-Saladin E, Wilson DL, Dickerson IM (1994) Isolation and in situ localization of a cDNA encoding a Kex2-like prohormone convertase in the nematode Caenorhabditis elegans. *Cell Mol Neurobiol* 14: 9–25.

[395] Sieburth D, Ch'ng Q, Dybbs M, Tavazoie M, Kennedy S, Wang D, Dupuy D, Rual JF, Hill DE, Vidal M, Ruvkun G, Kaplan JM (2005) Systematic analysis of genes required for synapse structure and function. *Nature* 436: 510–517.

[396] Husson SJ, Schoofs L (2007) Altered neuropeptide profile of Caenorhabditis elegans lacking the chaperone protein 7B2 as analyzed by mass spectrometry. *FEBS Lett* 581: 4288–4292.

[397] Morash MG, Soanes K, Anini Y (2011) Prohormone processing in zebrafish. *Methods Mol Biol* 768: 257–271.

[398] Leak TS, Keene KL, Langefeld CD, Gallagher CJ, Mychaleckyj JC, Freedman BI, Bowden DW, Rich SS, Sale MM (2007) Association of the proprotein convertase subtilisin/kexin-type 2 (PCSK2) gene with type 2 diabetes in an African American population. *Mol Genet Metab* 92: 145–150.

[399] Fujimaki T, Kato K, Yokoi K, Oguri M, Yoshida T, Watanabe S, Metoki N, Yoshida H, Satoh K, Aoyagi Y, Nozawa Y, Kimura G, Yamada Y (2010) Association of genetic variants

in SEMA3F, CLEC16A, LAMA3, and PCSK2 with myocardial infarction in Japanese individuals. *Atherosclerosis* 210: 468–473.

[400] Bouatia-Naji N, Vatin V, Lecoeur C, Heude B, Proenca C, Veslot J, Jouret B, Tichet J, Charpentier G, Marre M, Balkau B, Froguel P, Meyre D (2007) Secretory granule neuro-endocrine protein 1 (SGNE1) genetic variation and glucose intolerance in severe childhood and adult obesity. *BMC Med Genet* 8: 44.

[401] Marzban L, Rhodes CJ, Steiner DF, Haataja L, Halban PA, Verchere CB (2006) Impaired NH2-terminal processing of human proislet amyloid polypeptide by the prohormone convertase PC2 leads to amyloid formation and cell death. *Diabetes* 55: 2192–2201.

[402] Anini Y, Mayne J, Gagnon J, Sherbafi J, Chen A, Kaefer N, Chretien M, Mbikay M (2010) Genetic deficiency for proprotein convertase subtilisin/kexin type 2 in mice is associated with decreased adiposity and protection from dietary fat-induced body weight gain. *Int J Obes (Lond)* 34: 1599–1607.

[403] Chiu S, Kim K, Haus KA, Espinal GM, Millon LV, Warden CH (2007) Identification of positional candidate genes for body weight and adiposity in subcongenic mice. *Physiol Genomics* 31: 75–85.

[404] Iino K, Oki Y, Yamashita M, Matsushita F, Hayashi C, Yogo K, Nishizawa S, Yamada S, Maekawa M, Sasano H, Nakamura H (2010) Possible relevance between prohormone convertase 2 expression and tumor growth in human adrenocorticotropin-producing pituitary adenoma. *J Clin Endocrinol Metab* 95: 4003–4011.

[405] Tateno T, Izumiyama H, Doi M, Yoshimoto T, Shichiri M, Inoshita N, Oyama K, Yamada S, Hirata Y (2007) Differential gene expression in ACTH -secreting and non-functioning pituitary tumors. *Eur J Endocrinol* 157: 717–724.

[406] Hashimoto K, Koga M, Kouhara H, Arita N, Hayakawa T, Kishimoto T, Sato B (1994) Expression patterns of messenger ribonucleic acids encoding prohormone convertases (PC2 and PC3) in human pituitary adenomas. *Clin Endocrinol (Oxf)* 41: 185–191.

[407] Tani Y, Sugiyama T, Izumiyama H, Yoshimoto T, Yamada S, Hirata Y (2011) Differential gene expression profiles of POMC-related enzymes, transcription factors and receptors between non-pituitary and pituitary ACTH-secreting tumors. *Endocr J* 58: 297–303.

[408] Du J, Keegan BP, North WG (2001) Key peptide processing enzymes are expressed by breast cancer cells. *Cancer Lett* 165: 211–218.

[409] North WG, Du J (1998) Key peptide processing enzymes are expressed by a variant form of small-cell carcinoma of the lung. *Peptides* 19: 1743–1747.

[410] Kajiwara H, Itoh Y, Itoh J, Yasuda M, Osamura RY (1999) Immunohistochemical expressions of prohormone convertase (PC)1/3 and PC2 in carcinoids of various organs. *Tokai J Exp Clin Med* 24: 13–20.

[411]   Cheng M, Watson PH, Paterson JA, Seidah N, Chretien M, Shiu RP (1997) Pro-protein convertase gene expression in human breast cancer. *Int J Cancer* 71: 966–971.

[412]   Konoshita T, Gasc JM, Villard E, Takeda R, Seidah NG, Corvol P, Pinet F (1994) Expression of PC2 and PC1/PC3 in human pheochromocytomas. *Mol Cell Endocrinol* 99: 307–314.

[413]   Konoshita T, Gasc JM, Villard E, Seidah NG, Corvol P, Pinet F (1994) Co-expression of PC2 and proenkephalin in human tumoral adrenal medullary tissues. *Biochimie* 76: 241–244.

[414]   Guillemot J, Barbier L, Thouennon E, Vallet-Erdtmann V, Montero-Hadjadje M, Lefebvre H, Klein M, Muresan M, Plouin PF, Seidah N, Vaudry H, Anouar Y, Yon L (2006) Expression and processing of the neuroendocrine protein secretogranin II in benign and malignant pheochromocytomas. *Ann N Y Acad Sci* 1073: 527–532.

[415]   Kimura N, Pilichowska M, Okamoto H, Kimura I, Aunis D (2000) Immunohistochemical expression of chromogranins A and B, prohormone convertases 2 and 3, and amidating enzyme in carcinoid tumors and pancreatic endocrine tumors. *Mod Pathol* 13: 140–146.

[416]   Prabakaran I, Grau JR, Lewis R, Fraker DL, Guvakova MA (2011) Rap2A Is Upregulated in Invasive Cells Dissected from Follicular Thyroid Cancer. *J Thyroid Res* 2011: 979840.

[417]   Kimura N, Ishikawa T, Sasaki Y, Sasano N, Onodera K, Shimizu Y, Kimura I, Steiner DF, Nagura H (1996) Expression of prohormone convertase, PC2, in adrenocorticotropin-producing thymic carcinoid with elevated plasma corticotropin-releasing hormone. *J Clin Endocrinol Metab* 81: 390–395.

[418]   Mbikay M, Seidah NG, M C (2001) Neuroendocrine secretory protein 7B2: structure, expression and functions. *Biochem J* 357: 329–342.

[419]   Stridsberg M, Eriksson B, Janson ET (2008) Measurements of secretogranins II, III, V and proconvertases 1/3 and 2 in plasma from patients with neuroendocrine tumours. *Regul Pept* 148: 95–98.

[420]   Jaeger E, Webb E, Howarth K, Carvajal-Carmona L, Rowan A, *et al.* (2008) Common genetic variants at the CRAC1 (HMPS) locus on chromosome 15q13.3 influence colorectal cancer risk. *Nat Genet* 40: 26–28.

[421]   Prigge ST, Kolhekar AS, Eipper BA, Mains RE, Amzel LM (1997) Amidation of bioactive peptides: the structure of peptidylglycine alpha-hydroxylating monooxygenase. *Science* 278: 1300–1305.

[422]   Lee YC, Damholt AB, Billestrup N, Kisbye T, Galante P, Michelsen B, Kofod H, Nielsen JH (1999) Developmental expression of proprotein convertase 1/3 in the rat. *Mol Cell Endocrinol* 155: 27–35.

[423]   Creemers JW, Pritchard LE, Gyte A, Le Rouzic P, Meulemans S, *et al.* (2006) Agouti-related protein is posttranslationally cleaved by proprotein convertase 1 to generate agouti-

related protein (AGRP)83–132: interaction between AGRP83–132 and melanocortin receptors cannot be influenced by syndecan-3. *Endocrinology* 147: 1621–1631.

[424]   Doblinger A, Becker A, Seidah NG, Laslop A (2003) Proteolytic processing of chromogranin A by the prohormone convertase PC2. *Regul Pept* 111: 111–116.

[425]   Udupi V, Lee HM, Kurosky A, Greeley GH, Jr. (1999) Prohormone convertase-1 is essential for conversion of chromogranin A to pancreastatin. *Regul Pept* 83: 123–127.

[426]   Eskeland NL, Zhou A, Dinh TQ, Wu H, Parmer RJ, Mains RE, O'Connor DT (1996) Chromogranin A processing and secretion: specific role of endogenous and exogenous prohormone convertases in the regulated secretory pathway. *J Clin Invest* 98: 148–156.

[427]   Arden SD, Rutherford NG, Guest PC, Curry WJ, Bailyes EM, Johnston CJ, Hutton JC (1994) The posttranslational processing of chromogranin A in the pancreatic islet: involvement of the eukaryote subtilisin PC2. *Biochem J* 298: 521–528.

[428]   Laslop. A, Weiss C, Savaria D, Eiter C, Tooze SA, Seidah NG, Winkler H (1998) Proteolytic processing of chromogranin B and secretogranin II by prohormone convertases. *J Neurochem* 70: 374–383.

[429]   Dey A, Xhu X, Carroll R, Turck CW, Stein J, Steiner DF (2003) Biological processing of the CART precursor by prohormone convertases PC2 and PC1/PC3. *J Biol Chem* 278: 15007–15014.

[430]   Basak A, Ernst B, Brewer D, Seidah NG, Munzer JS, Lazure C, Lajoie GA (1997) Histidine-rich human salivary peptides are inhibitors of proprotein convertases furin and PC7 but act as substrates for PC1. *J Pept Res* 49: 596–603.

[431]   Laslop A, Doblinger A, Weiss U (2000) Proteolytic processing of chromogranins. *Adv Exp Med Biol* 482: 155–166.

[432]   Rosenblatt MI, Dickerson IM (1997) Endoproteolysis at tetrabasic amino acid sites in procalcitonin gene-related peptide by pituitary cell lines. *Peptides* 18: 567–576.

[433]   Wang W, Beinfeld MC (1997) Cleavage of CCK 33 by recombinant PC2 *in vitro*. *Biochem Biophys Res Commun* 231: 149–152.

[434]   Wang W, Birch NP, Beinfeld MC (1998) Prohormone convertase 1 (PC1) when expressed with pro cholecystokinin (pro CCK) in L cells performs three endoproteolytic cleavages which are observed in rat brain and in CCK-expressing endocrine cells in culture, including the production of glycine and arginine extended CCK8. *Biochem Biophys Res Commun* 248: 538–541.

[435]   Vishnuvardhan D, Connolly K, Cain B, Beinfeld MC (2000) PC2 and 7B2 null mice demonstrate that PC2 is essential for normal pro-CCK processing. *Biochem Biophys Res Commun* 273: 188–191.

[436]   Tagen MB, Beinfeld MC (2005) Recombinant prohormone convertase 1 and 2 cleave

purified pro cholecystokinin (CCK) and a synthetic peptide containing CCK 8 Gly Arg Arg and the carboxyl-terminal flanking peptide. *Peptides* 26: 2530–2535.

[437] Brar B, Sanderson T, Wang N, Lowry PJ (1997) Post-translational processing of human procorticotrophin-releasing factor in transfected mouse neuroblastoma and Chinese hamster ovary cell lines. *J Endocrinol* 154: 431–440.

[438] Miller R, Toneff T, Vishnuvardhan D, Beinfeld M, Hook VY (2003) Selective roles for the PC2 processing enzyme in the regulation of peptide neurotransmitter levels in brain and peripheral neuroendocrine tissues of PC2 deficient mice. *Neuropeptides* 37: 140–148.

[439] Johanning K, Mathis JP, Lindberg I (1996) Role of PC2 in proenkephalin processing: antisense and overexpression studies. *J Neurochem* 66: 898–907.

[440] Breslin MB, Lindberg I, Benjannet S, Mathis JP, Lazure C, Seidah NG (1993) Differential processing of proenkephalin by prohormone convertases 1(3) and 2 and furin. *J Biol Chem* 268: 27084–27093.

[441] Dickinson CJ, Sawada M, Guo YJ, Finniss S, Yamada T (1995) Specificity of prohormone convertase endoproteolysis of progastrin in AtT-20 cells. *J Clin Invest* 96: 1425–1431.

[442] Sawada M, Finniss S, Dickinson CJ (2000) Diminished prohormone convertase 3 expression (PC1/PC3) inhibits progastrin post-translational processing. *Regul Pept* 89: 19–28.

[443] Bramante G, Patel O, Shulkes A, Baldwin GS (2011) Ferric ions inhibit proteolytic processing of progastrin. *Biochem Biophys Res Commun* 404: 1083–1087.

[444] Rehfeld JF, Zhu X, Norrbom C, Bundgaard JR, Johnsen AH, Nielsen JE, Vikesaa J, Stein J, Dey A, Steiner DF, Friis-Hansen L (2008) Prohormone convertases 1/3 and 2 together orchestrate the site-specific cleavages of progastrin to release gastrin-34 and gastrin-17. *Biochem J* 415: 35–43.

[445] Zhu X, Cao Y, Voogd K, Steiner DF (2006) On the processing of proghrelin to ghrelin. *J Biol Chem* 281: 38867–38870.

[446] Dhanvantari S, Seidah NG, Brubaker PL (1996) Role of prohormone convertases in the tissue-specific processing of proglucagon. *Mol Endocrinol* 10: 342–355.

[447] Rouille Y, Bianchi M, Irminger JC, Halban PA (1997) Role of prohormone convertase PC2 in the processing of proglucagon to glucagon. *FEBS Lett* 413: 119–123.

[448] Rouille Y, Kantengwa S, Irminger JC, Halban PA (1997) Role of the prohormone convertases in the processing of proglucagon to glucagon-like peptides. *J Biol Chem* 72: 32810–32816.

[449] Dhanvantari S, Brubaker PL (1998) Proglucagon processing in an islet cell line: effects of PC1 overexpression and PC2 depletion. *Endocrinol* 139: 1630–1637.

[450] Rothenberg ME, Eilertson CD, Klein K, Zhou Y, Lindberg I, McDonald JK, Mackin RB, Noe BD (1995) Processing of mouse proglucagon by recombinant prohormone convertase 1 and immunopurified prohormone convertase 2 in vitro. *J Biol Chem* 270: 10136–10146.

[451] Irminger JC, Meyer K, Halban P (1996) Proinsulin processing in the rat insulinoma cell line INS after overexpression of the endoproteases PC2 or PC3 by recombinant adenovirus. *Biochem J* 320 (Pt 1): 11–15.

[452] Kaufmann JE, Irminger JC, Mungall J, Halban PA (1997) Proinsulin conversion in GH3 cells after coexpression of human proinsulin with the endoproteases PC2 and/or PC3. *Diabetes* 46: 978–982.

[453] Higham CE, Hull RL, Lawrie L, Shennan KI, Morris JF, Birch NP, Docherty K, Clark A (2000) Processing of synthetic pro-islet amyloid polypeptide (proIAPP) 'amylin' by recombinant prohormone convertase enzymes, PC2 and PC3, in vitro. *Eur J Biochem* 267: 4998–5004.

[454] Viale A, Ortola C, Hervieu G, Furuta M, Barbero P, Steiner DF, Seidah NG, Nahon JL (1999) Cellular localization and role of prohormone convertases in the processing of pro-melanin concentrating hormone in mammals. *J Biol Chem* 274: 6536–6545.

[455] Brakch N, Rist B, Beck-Sickinger AG, Goenaga J, Wittek R, Burger E, Brunner HR, Grouzmann E (1997) Role of prohormone convertases in pro-neuropeptide Y processing: coexpression and in vitro kinetic investigations. *Biochemistry* 36: 16309–16320.

[456] Rovere C, Barbero P, Kitabgi P (1996) Evidence that PC2 is the endogenous pro-neurotensin convertase in rMTC 6–23 cells and that PC1- and PC2-transfected PC12 cells differentially process pro-neurotensin. *J Biol Chem* 271: 11368–11375.

[457] Villeneuve P, Feliciangeli S, Croissandeau G, Seidah NG, Mbikay M, Kitabgi P, Beaudet A (2002) Altered processing of the neurotensin/neuromedin N precursor in PC2 knock down mice: a biochemical and immunohistochemical study. *J Neurochem* 82: 783–793.

[458] Benjannet S, Rondeau N, Day R, Chretien M, Seidah NG (1991) PC1 and PC2 are pro-protein convertases capable of cleaving proopiomelanocortin at distinct pairs of basic residues. *Proc Natl Acad Sci U S A* 88: 3564–3568.

[459] Allen R, Peng B, Pellegrino M, Miller E, Grandy D, Lundlblad J, Washburn C, Pintar J (2001) Altered processing of pro-orphanin FQ/Nociceptin and pro-opiomelanocortin-derived peptides in the brains of mice expressing defective prohormone convertase 2. *J Neurosci* 21: 5864–5870.

[460] Hardiman A, Friedman TC, Grunwald WC, Jr., Furuta M, Zhu Z, Steiner DF, Cool DR (2005) Endocrinomic profile of neurointermediate lobe pituitary prohormone processing in PC1/3- and PC2-Null mice using SELDI-TOF mass spectrometry. *J Mol Endocrinol* 34: 739–751.

[461] Galanopoulou AS, Kent G, Rabbani SN, Seidah NG, Patel YC (1993) Heterologous processing of prosomatostatin in constitutive and regulated secretory pathways. Putative role of the endoproteases furin, PC1, and PC2. *J Biol Chem* 268: 6041–6049.

[462]   Galanopoulou AS, Seidah NG, Patel YC (1995) Heterologous procesing of rat proso-matostatin to somatostatin-14 by PC2: requirement for secretory cell but not the secretion granule. *Biochem J* 311: 111–118.

[463]   Brakch N, Galanopoulou AS, Patel YC, Boileau G, Seidah NG (1995) Comparative pro-teolytic processing of rat prosomatostatin by the convertases PC1, PC2, furin, PACE4 and PC5 in constitutive and regulated secretory pathways. *FEBS Lett* 362: 143–146.

[464]   Toll L, Khroyan TV, Sonmez K, Ozawa A, Lindberg I, McLaughlin JP, Eans SO, Shahien AA, DR. K (2011) Peptides derived from the prohormone proNPQ/spexin are potent cen-tral modulators of cardiovascular and renal function and nociception. *FASEB J* in press.

[465]   Friedman TC, Loh YP, Cawley NX, Birch NP, Huang SS, Jackson IM, Nillni EA (1995) Processing of prothyrotropin-releasing hormone (Pro-TRH) by bovine intermediate lobe secretory vesicle membrane PC1 and PC2 enzymes. *Endocrinology* 136: 4462–4472.

[466]   Nillni EA, Friedman TC, Todd RB, Birch NP, Loh YP, Jackson IM (1995) Pro-thyrotropin-releasing hormone processing by recombinant PC1. *J Neurochem* 65: 2462–2472.

[467]   Schaner P, Todd RB, Seidah NG, Nillni EA (1997) Processing of prothyrotropin-releasing hormone by the family of prohormone convertases. *J Biol Chem* 272: 19958–19968.

[468]   Nillni EA (1999) Neuroregulation of ProTRH biosynthesis and processing. *Endocrine* 10: 185–199.

[469]   Gabreels BA, Swaab DF, de Kleijn DP, Dean A, Seidah NG, Van de Loo JW, Van de Ven WJ, Martens GJ, Van Leeuwen FW (1998) The vasopressin precursor is not processed in the hypothalamus of Wolfram syndrome patients with diabetes insipidus: evidence for the involvement of PC2 and 7B2. *J Clin Endocrinol Metab* 83: 4026–4033.

[470]   Trani E, Giorgi A, Canu N, Amadoro G, Rinaldi AM, Halban PA, Ferri GL, Possenti R, Schinina ME, Levi A (2002) Isolation and characterization of VGF peptides in rat brain. Role of PC1/3 and PC2 in the maturation of VGF precursor. *J Neurochem* 81: 565–574.

[471]   Urbe S, Dittie AS, Tooze SA (1997) pH-dependent processing of secretogranin II by the endopeptidase PC2 in isolated immature secreotry granules. *Biochem J* 321: 65–74.

[472]   Dittie AS, Tooze SA (1995) Characterization of the endopeptidasePC2 activity towards secretogranin II in stably transfected PC12 cells. *Biochem J* 310: 777–787.

Author Biographies

Akira Hoshino was born in Yokohama, Japan and moved to the United States to attend Vassar College where she earned her B.A. degree in Biology and Chinese. She is currently a graduate student in the Program in Neuroscience at the University of Maryland–Baltimore, working toward her Ph.D. under Dr. Iris Lindberg. Her main interests in the Lindberg lab are regulation of PC1/3 and the role of proSAAS in protein aggregation-mediated neurodegenerative diseases. She has shown that PC1/3 activity is regulated via self-association and that proSAAS can function as an anti-aggregant. In 2012, she will be heading to Seattle, Washington to research stem cell replacement therapies to treat retinal degeneration. In her spare time, she enjoys baking and doing Pilates.

Iris Lindberg (Ph.D. 1980 University of Wisconsin, Madison) has been working on prohormone convertases since 1991; her group, now at the University of Maryland–Baltimore, focuses on the enzymology and cell biology (and occasionally evolution) of these unusual serine proteinases and their binding proteins. Her group was the first to purify recombinant PC1/3 and PC2, and they are now working to crystallize these important enzymes. Her team also showed that 7B2 is required for proPC2 to mature to an active enzyme; 7B2 does so by blocking the spontaneous aggregation of proPC2. Recent work in the lab is centered on deriving new convertase inhibitors as well as exploring the idea that 7B2 and proSAAS represent general secretory chaperones active in neuro-degenerative disease.